second skin

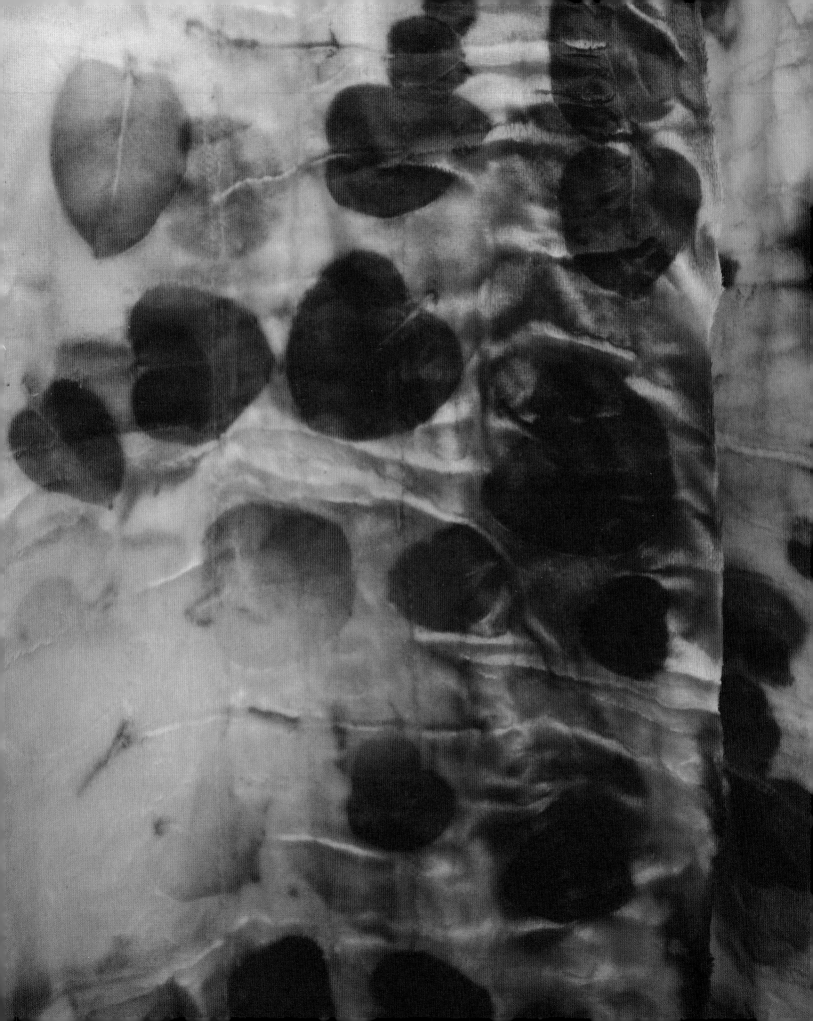

second skin
choosing and caring for textiles and clothing

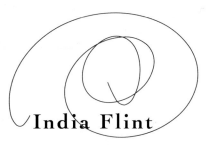
India Flint

MURDOCH BOOKS

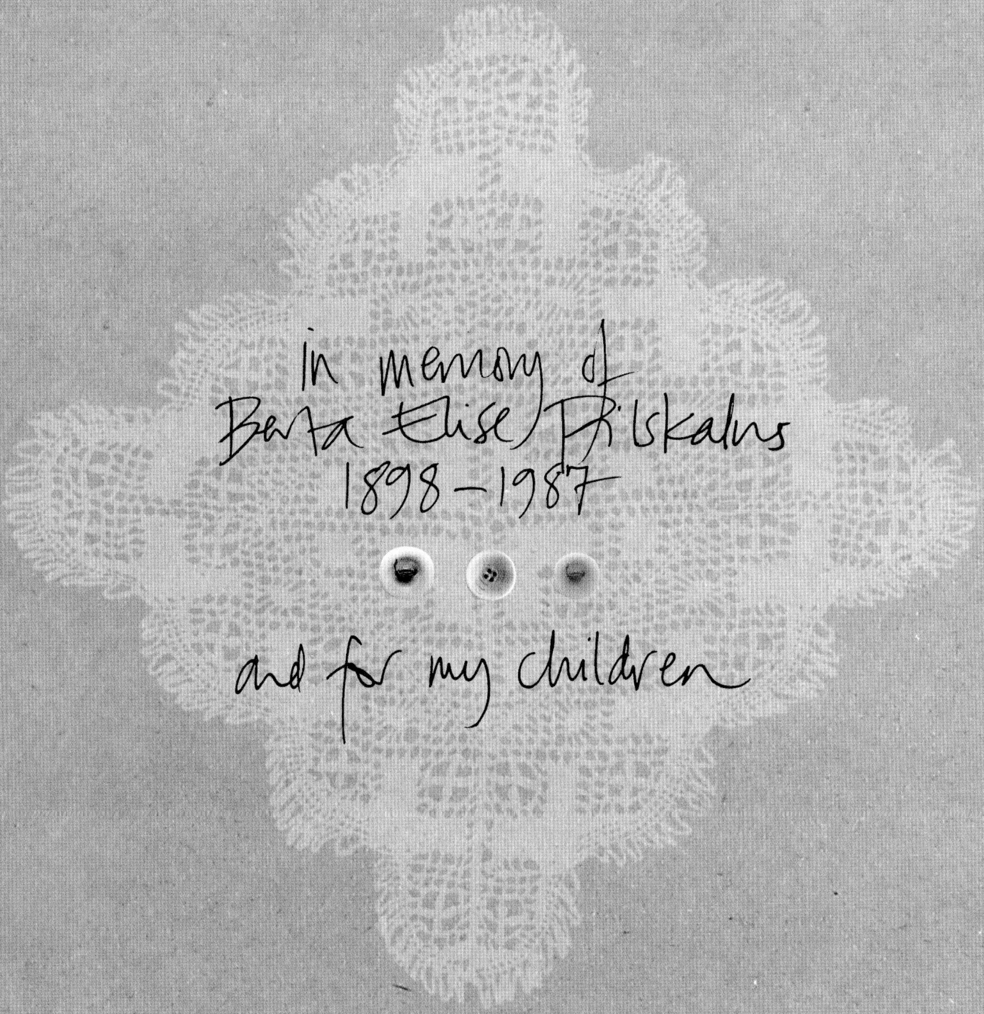

In memory of
Berta Elise Pilskalns
1898–1987

and for my children

a rectangle of cloth

to wrap the baby, make the bed,

grace the meal and honour the guest,

to mop up a spill, encircle a waist,

screen the window and admit the breeze,

to proclaim a cause,

to tend the corpse …

Gwen Egg[1]

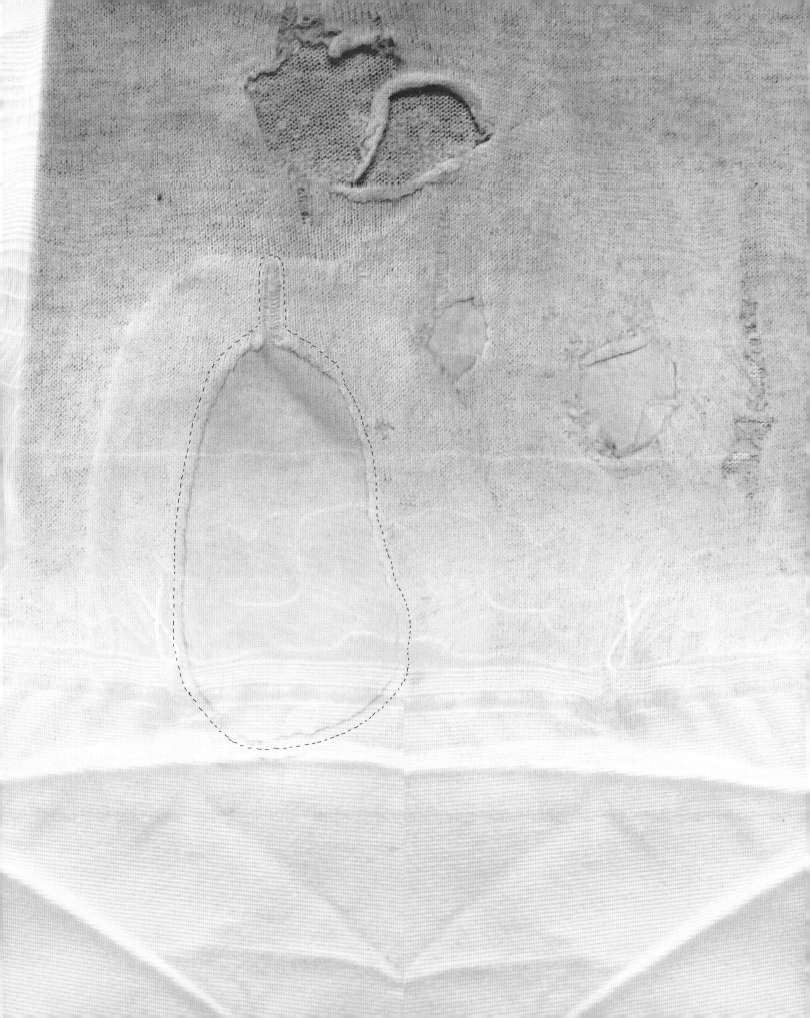

Contents

1 Preface

chapter 1
16 **Underpinnings**
on the role of cloth and clothing and
how fashion and consumption are affecting
the ecology

chapter 2
40 **Provenance**
a look at what textiles are made from and
some of their properties, with an emphasis on
those derived from natural sources

chapter 3
78 **Clothing choices**
making informed choices – some things
to think about when making decisions
about clothes

chapter 4
96 **Planning your wardrobe**
planning and managing the wardrobe; how
much clothing do we really need?

chapter 5
110 **Making clothes**
on making clothes, sourcing materials and the
satisfaction in wearing self-made clothing

chapter 6
134 Maintenance
keeping up appearances; maintenance, mending and making-do

chapter 7
162 Gallery
pictures worth a thousand words – material matters, metamorphoses and making

chapter 8
186 Repurposed and repurposing fashion
giving your second skin a second life – can it be tweaked or is major surgery necessary?

chapter 9
208 When all else fails
patching, piecing, felting and twining

chapter 10
234 Dyeing as if life depended on it
some simple and safe processes for colouring cloth with plants

chapter 11
268 And in the very end
the choice of raiment for the last dance of all

notes 274
useful websites 276
further reading 277
index 278
photo captions and credits 284
about the author 285
acknowledgments 286

Preface

It is clear that Earth is in crisis, as resources are consumed at unprecedented levels, food and water supplies are inadequate to meet potential demand and industries around the globe continue to pump out pollutants with generally scant regard for the ecology. Ignoring the advice of our forebears we tread heavily on the land, leaving ineradicably deep footprints. If we consider Earth as one large organism it is apparent that anything we do to our bio-region not only affects the rest of the planet but also eventually comes back to ourselves. Simplify this to the view that the world is a community to which we all belong and our responsibilities become very clear. Many of our day-to-day activities have quite appalling consequences and sometimes when natural disaster strikes it would be easy to think that our Mother Earth, like a flea-infested dog, is doing her best to rid herself of us annoying parasites. It's become a matter of changing … or perishing.

Listening to the forecasts of the doomsayers it can be hard to see how we as individuals can make a difference to the situation. Some say global warming is on our doorstep, other scientists suggest the pendulum could swing as easily to a new ice age. It would be all too easy to begin to feel like Frodo Baggins[2] when faced with the task of destroying the One Ring, that the road ahead is dark and confusing and quite likely overwhelming.

We may not all have the financial resources to install solar panels for electricity or hot water, or to harness wind energy for our homes, yet we can nonetheless choose paths that will do good and hope (as the French painter Henri Matisse is reputed to have done) that 'there are always flowers for those who want to see them'.

Each journey begins with a single step.

This book is not about moving mountains but about some of the more easily achievable things we might do for our planet by living simpler lives and using less of the world's non-renewable resources, specifically those to do with cloth and clothing, in a more considered and informed manner. I hope that the advice in it will be useful when choosing fibres and deciding on their attendant maintenance with regard to effects on human health as well as that of the environment, and in making the clothing that acts as our second skin last as long as possible.

There may even be a few contradictions as various ideas are presented. What you do with it is up to you. *Second Skin* does not pretend to be a textbook or an academic tome; my aspiration was rather to present information that could be of use together with interesting work by other artists and designers who are thinking about similar challenges, as well as the occasional small project that could be carried out by the individual using resources readily available at home or in the garden. I also trust that readers will indulge me in the inclusion of small personal stories along the way, such as this one about my grandmother, who was left motherless not long after she came into this world.

felted together from pre

Grandmother was farmed out to be cared for by others and consequently did not have as happy a childhood as she might have done had her mother survived. Be that as it may, one Christmas Eve at the local vicarage, when presents were being handed out in the traditional manner – one for each child – Grandmother was handed a large ball of wool, together with a small crochet hook. Being an industrious child she set about crocheting a scarf, and as she worked her way through the ball of yarn, surprises began to appear – first a hairclip, then a ribbon, further in a small comb. Gradually as the scarf grew, so did the collection of pretty objects that had been carefully wound into a beautiful bundle of surprises by the pastor's two daughters (community-minded women who ran the village school), not wishing to single out the motherless child at the moment of bestowal by giving her more than the others; clearly aware of Grandmother's interest in making things but also thoughtfully extending her pleasure in the gift of wool and hook with the unexpected appearance of treasures.

In effect, I'm tossing a stone into the water and watching where the ripples go. I hope you find a treat or three as well as the odd useful idea tucked between these pages.

chapter 1
Underpinnings

Be the change you wish to see in the world.   Mahatma Gandhi

A life in cloth

Almost from the moment of our birth we are wrapped in textiles of one form or another. Children of my vintage were received into a cloth on emerging into the world, so instead of having skin contact with our mother as our first sensory experience, the scent of some piece of fabric, usually a utilitarian hospital textile, will be embedded in our memory. While in some cultures births are celebrated with a specially embellished textile hand-worked for the purpose, I suspect most readers of this book will not have been so blessed.

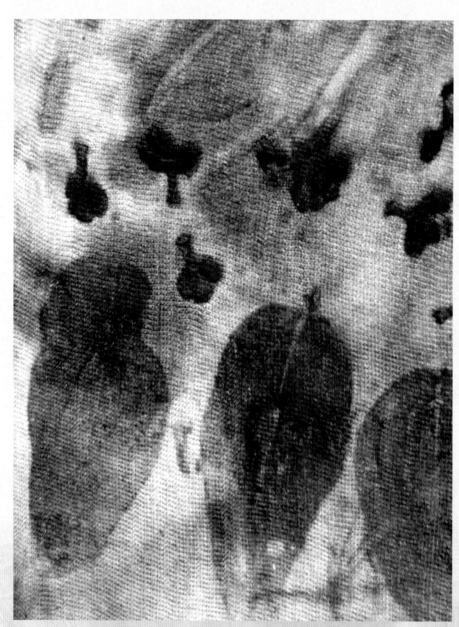

fragment from a discarded merino jersey used as a cleaning rag and subsequently revived with plant dyes

We have a fundamental need for clothing; it forms our second skin, protecting us from the elements, preserving modesty, satisfying the basic human desire for decoration and sending distinct social messages about us to other people. Just as the extremes of fashion denote our social aspirations and personal preferences, so military or work-related uniforms signal our occupations. Specific religious attire such as the habits of Buddhist or Christian monks and nuns, the yarmulke of the Jewish people, the bonnets of the Amish and the enveloping drapes espoused by women of the Muslim faith send very clear messages about the beliefs of the wearers. Wandering through a marketplace in the company of my aunt, a Lutheran deaconess, when she happens to be wearing full regalia is a completely different experience to being with the same aunt in the same place when she is dressed in mufti. It is so much easier to change our personal appearance by switching clothes than to change actual aspects of our physical incarnation.

tattoo artist Roger Ingerton (New Zealand) and a happy client inspecting the results of his labours

In many cultures, especially those in which clothing plays a lesser role (perhaps due to gentler climates), the skin itself might be decorated by body paint, scarification or tattooing. Once regarded principally as the province of sailors and criminals by those in Western cultures, tattoos are gaining in popularity and propriety. Respectable matrons flaunting butterflies and proud fathers bearing the names of their offspring have joined those who once sought exclusivity by having such permanent decorations applied to their skins.

Tourists collect tattoos as holiday souvenirs, often misappropriating symbols belonging to cultures that are not their own. The cheerful, haphazard and often improper borrowing of Maori *moko* springs to mind. Traditionally the incising of *ta moko* on the skin was preceded by months of consultation with elders and confirmed the person's commitment to their Maori heritage rather than being a souvenir of New Zealand or simply an unplanned acquisition during a tipsy night out on the town. Such adornments are not to be taken lightly, being rather difficult to remove if one tires of them, nor are they considered universally acceptable; so using clothing choices to signal affiliations while fulfilling the practical functions of protection and propriety remains of some consequence in our society.

traditional Japanese and contemporary Western-influenced Japanese dress meet on a street in Yamaguchi

The notion of dressing appropriately varies according to our cultural heritage. A self-respecting Frenchwoman wouldn't dream of being seen in a pair of trackpants (the reasoning being that sloppy dressing is an offence to others), whereas in Australia such concerns are largely shrugged off. The question of how we choose our clothing is important, not only in terms of the picture we wish to present to the rest of the world but also in regard to the impact that our choices have on the ecology.

◀ first skin marked with a tattoo, the second coloured with eucalyptus

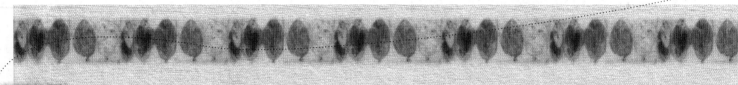

Nothing that we do is without effect;

however, back in the days when Earth's population was far smaller, our collectively inflicted damage was rather less noticeable. In the twenty-first century the world is burdened with so many people and the population is growing at such a rate that humankind is having an unprecedented effect on the natural systems of our world. Micro-particle pollution from coal-burning energy industries is affecting rainfall patterns across Australia. Mountainous regions that once boasted good rains now receive far less precipitation as water molecules bond with small smoke particles, forming electrically charged clusters that are too small to form actual raindrops, so the water vapour is carried off elsewhere by prevailing winds. Emissions fill the atmosphere with filth and affect air quality. Governments protest that carbon is the culprit and propose extra taxes while continuing to clear-fell old-growth forests and increasing the output of coalmines.

Trees and forests are one of the simplest and most practical means of storing carbon. They are also our lungs, converting carbon dioxide to oxygen in the process of photosynthesis. Coal is compressed carbon-rich matter and leaving it in the earth would be such a simple solution to the perceived problem. Despite the vilification of this element it must be remembered that carbon is actually vital to life. The study of organic chemistry (that is, the chemistry of living things) is, after all, the study of compounds containing carbon. Unfortunately our societies have become completely dependent on generated power and the choice between coal and nuclear energy is very similar to that between the devil and the deep blue sea. Nuclear power stations require vast amounts of energy in their construction and produce a toxic product that is impossible to store safely, takes millions of years to break down and gives off dangerous radiation while it does so. Using coal power wisely to produce energy that is in turn used to create the technology required to harvest sustainable energies such as wind, solar and wave power would seem to be the sensible option. Sadly, however, this is unlikely to occur while industry leaders and governments still measure success in terms of dollars and 'growth'. The industrial notions of 'growth' and 'development' and the greed for profits that drive our modern world will be its downfall.

THIS GARMENT HOLDS A STORY. IT HIDES IN THE SEAMS AND WHISPERS FROM THE FOLDS. LOOK CLOSE: A COTTON FIELD BAKES IN THE SUN, A PICKER CARRIES A BULGING SACK, A SEWING MACHINE HUMS. TRACE EACH STITCH BACK TO HANDS LIKE YOURS...

IT IS TIME TO CHANGE OUR CLOTHES

CARE INSTRUCTIONS ON REVERSE

How is all this relevant to clothing? The shocking facts are that the industrial manufacture of clothing is rife with excess and wastage. More clothes are produced than are needed and the off-cuts simply dumped in landfills (where they will eventually be joined by the unsold surplus, unless it has been burned instead). The materials required for manufacture are transported around the world not only to reach the eventual point of sale but also in the process of making. A woollen jumper purchased in Australia may be made from Australian-grown wool that is likely to have been scoured in China, spun in India and then machine knitted from that intermediate product in yet another (probably Asian) country before returning to the country of origin. It is very difficult to establish a reliable processing chain for the raw material in Australia as it is considered to be more economical to exploit lower wages and less stringent environmental regulations elsewhere. The processes and treatments taken for granted as necessary in textiles and clothing production are (for me) the stuff of nightmares; in order for fungicides, dyes and stain-resistant treatments to be applied to cloth, the substances themselves need to be produced somewhere. Begin to examine one part of the process and you'll find an ever-growing network of related environmental (and social) impacts, metaphorically writhing away like an angry octopus, leading you on a complex trail like a constantly expanding Mandelbrot set. If consumers really knew what they were wearing they would be appalled. I wish I'd known more when I was dressing my babies. Making clothing requires more than just cloth and dyes; buttons, zips and all sorts of other trims are manufactured as well as swing-tags, packaging, plastic coat-hangers … the list is endless. In Britain, for example, more than two thirds of textiles used are man-made (synthetic).[3] With the exception of cloth derived from bamboo, wood or other synthesised plant substances, those fabrics have their origins in oil. Whereas cars and their drivers are usually vilified for their use of fossil fuels, the making of plastics and synthetic fabrics eclipses the demands of motoring by far. Australian fashion writer Lou Pardi succinctly sums the problem thus: 'Modern production of clothing is an environmental and social disaster.'[4]

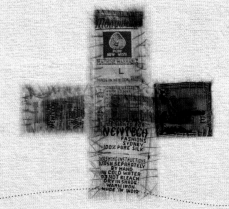

It seems to me that many of us developed countries are seeking to fulfil ourselves through acquisitions, needing the newest and the best like heroin junkies in search of a bigger and better thrill. While the powerful and wealthy classes have a history of consumption coupled with a certain flamboyance, until about the middle of the last century the working-class person only owned the clothes they needed to get through life and thought very carefully before making or purchasing something new. Somewhere in the last fifty years shopping became a recreational activity and fashion became something directed by manufacturers seeking a profit rather than having its roots in the quirks of some conspicuous individual such as Beau Brummel or George, Lord Byron.

We might no longer adopt a dandelion in our buttonhole, apeing some public figure, but many of us do put effort into seeking out contemporary styles, not wishing to appear old-fashioned. We have become puppets to trends that are decided years in advance of production, while the discerning of these supposed trends has become an industry in itself. Fashion magazines collude with fashion manufacturers to highlight those seasonal aspects they particularly wish to promote, and although most people would like to be recognised for their individuality it is a rare person indeed who develops a personal aesthetic and dares to be radically different from the mainstream. Acceptance in society generally necessitates some sort of cultural conformity. This conventionality in turn requires the consumer to change their personal wardrobe in line with current trends. Once, a person might have purchased a coat as a purely serviceable garment with a view to having it last at least ten years and would have paid a price based on the quality of the fabric and the durability of the making; now the consumer buys a cheap garment acknowledging the intent to discard it at the end of the season.

sacred cow grazing on detritus in India

Consequently shopper demand for volumes of cheap textiles (among other commodities) has led to greater production output using supposedly economical but frequently exploitative methods, creating appalling situations for textile workers and leading to the world environment becoming severely compromised. Just as consumers have come to rely on the easy availability of take-away foods, so too have fast fashion outlets grown rapidly in popularity and number, feeding a perceived demand.

While as recently as the 1950s the corner dressmaker or bespoke tailor and a limited number of department and clothing stores provided those garments that were beyond the skill of the home dressmaker, these institutions became a rarity in the late twentieth century. Consumers are accustomed to acquiring the ready-made and those who make their own are now the exception rather than the rule. Fashion writer Colin MacDowell[5] roundly denounced the Western world some years ago, arguing that we 'drive big cars, own too many clothes, wash when there is no dirt and eat international foods that have travelled the globe'.[6] But he still participates actively in this world and derives a comfortable living from critiquing modern life. It would be very difficult not to, unless one decided to become a hermit in the Gobi Desert.

Fashion relies on perceived obsolescence

Fast fashion (the industry sector made most visible by the number of high-volume outlets in shopping precincts) encourages excessive consumption and the consumer's acquisition of far more than is necessary for basic survival, simply because to maintain profit levels the brand must constantly offer something new to solicit more purchases. Obsolescence is often deliberately built into the garment through the selection of materials and quality of workmanship as well as blatantly through the styling. Shopping has become an activity in which many consumers participate because they are seeking a leisure interest, not because they actually require specific commodities. People have become complacent about buying a cheap garment, wearing it for a few weeks and then throwing it away.

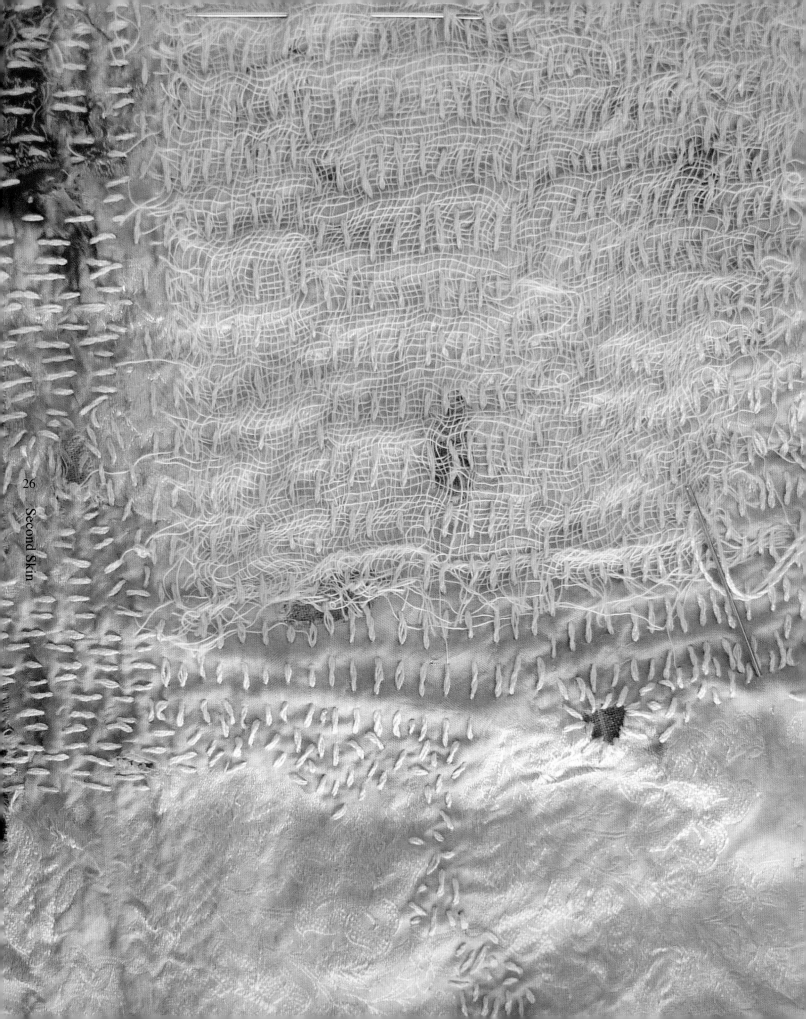

As a consequence thousands of tonnes of perfectly serviceable textile goods are annually dumped in landfills, lost in closets or transported (using even more fossil fuels) to markets in developing countries. While the availability of pre-used goods in such countries might well fill a need for cheap clothing, it also has the effect of globalising culture and nibbling away at regional dress styles. A young person in Dhaka may find himself looking much the same as his counterpart in New York or Sydney.

the warped-looking 'witch's broom' growth on this *Echium plantageneum* is evidence of the drift of glyphosate-based weedkiller sprays on the breeze. The sad thing is that this plant is in the middle of a national park in the Flinders Ranges

Mass production of rapidly constructed clothing made for the high street from often inferior materials coloured with synthetic dyes usually generates garments destined for pathetically short lives.

For example a cheap cotton T-shirt might have its origins in a cotton field in Asia, where the crop has been sprayed with various poisons to kill insects and reduce weed growth and subsequently is picked and bagged by field hands who are thus exposed to the chemicals during the harvest as well as during the application. Whole families of workers have become ill due to the cotton industry. Whether the problem is the fine fibres affecting lungs or the toxic chemicals workers are required to spray, or the conditions in the cotton mills where still more chemicals are needed to transform the fibre into something soft enough to wear, cotton is by no means a friendly fibre.

◀ handstitching discarded fragments into a 'new' composite piece

To suddenly put a stop to all this would be to deprive hundreds of thousands of people of a livelihood, but to continue to support the exploitation of people and the pollution of their lands by supporting fast fashion is frankly immoral. This kind of industry is directly responsible for damaging rivers, ground water and surrounding agricultural land in developing countries as well as polluting air (which travels across borders). Often, child labour is used for unskilled jobs in unhygienic conditions and hazardous occupations, with children working long hours for meagre wages. This places them in vulnerable situations and denies them their enjoyment of childhood as well as their right to education. Huge volumes of precious water are diverted to textile industries for dyeing, processing and rinsing cloth. Ground and surface waters are in turn polluted by effluent from these activities.

At the other extreme of the fashion world, a garment produced by an haute couture fashion house is carefully constructed, documented and ultimately treasured by the owner even though, ironically, the item might only be worn a few times before being packed away. Sometimes such garments find their way into museum collections, others eventually into vintage sales. Similarly, having one's clothing tailor-made or bespoke will ensure proper fit and careful making as well as the potential for remaking. Establishing a relationship with a tailor or seamstress makes good sense if you don't have a sewing gene.

Even when garments are made with great care and attention, most of the fabrics used still come from developing countries, where the textile industry is often under-regulated and may retain a significant environmental footprint.

This all sounds very much like doom and gloom, but it doesn't have to be that way.

People who are concerned for the direction in which the world is heading will wish to consider carefully where their wardrobe comes from. An early Greens senator in 1980s West Germany distinguished himself publicly by wearing only clothes from second-hand shops. In this instance the wearing of recycled and even darned goods was recognised as a consciously praiseworthy act.

Small measures can wander far

The clothes we wear are in constant contact with our most absorptive and largest organ. It makes sense to consider very carefully what we put on to our bodies. When labels state that colours may run you can be very sure that small particles of indeterminate chemical composition are going to end up on (and eventually be absorbed by) your skin. I wouldn't mind blue stains from naturally fermented indigo on my body but I do worry about potentially toxic stuff sneaking its way unnecessarily into my system. I choose to wear only clothing made from natural fibres. These fibres are either undyed or dyed naturally (usually by me). I do admittedly wear blue denim jeans when on horseback or working with sheep but they've all been sourced from op-shops and thrift stores, so previous wearers have washed out any excess dye that might have been caked on the surface of the cloth. It isn't merely dyes we need to worry about; stain repellents, water-proofing and fungicidal treatments might also have been applied to the cloth and are unlikely to have been listed on the label. If clothes, like foods, were required to list the ingredients (both chemical and social) that were used in the production it would make for horrific reading.

a woman repairing a garment, pictured on a Japanese ceramic fragment

I am fortunate to come from a long line of needlewomen and to have had the example of a mother, grandmother and aunts who all made their own clothes as well as what was necessary for their families. When my grandmother left Latvia in the closing years of World War II she carried with her the hand-cranked Singer sewing machine she had acquired in 1927, a quite amazing feat since she was a small woman and the thing is extremely heavy. This machine not only financially supported her family in the ensuing years, it was used to make clothes for several generations of children, many of whom also enjoyed their own first sewing exploits on it. I consider myself very lucky to have been endowed with the trusty Singer when my grandmother passed on to the Next Big Adventure.

Grandmother's 1927 Singer sewing machine, still in use today

In our family tradition (as in so many others), worn sheets were turned 'sides to middle' to extend their life, and when they became really fragile in parts, the best pieces were transformed into soft nightgowns while the tattier bits became dusters and cleaning cloths or were doubled a few times to act as liners for babies' nappies. Worn or torn garments were initially mended but eventually incorporated into patchwork blankets or ripped up to weave rag rugs (shredding sheets for rug making was an excellent way to vent frustration!). Socks were assiduously darned, as were our jumpers. Our favourite pants had impressive layers of patches. Grandmother's philosophy was that you needn't be embarrassed by mending.

During our 'growing' years my mother would proudly knit the most amazing jumpers for us each winter. She would devise some intricate pattern such as a proper six-pointed snowflake, undertake relatively complex calculations to ensure the image didn't warp in the knitting process and then produce a series of jumpers for us in various colour combinations so that the family would be wearing a harmonious story on the ski fields. She also sewed all of our clothes, dressing me in garments made from the remnants left after cutting her own outfits. My brother had a tiny Sunday suit made to match that of my father and both wore bow-ties made from the same fabric as our dresses. While we were little we were apparently quite happy to resemble a picture-book family but inevitably as we became teenagers the burning desire for a pair of jeans set in, much to my parents' disappointment.

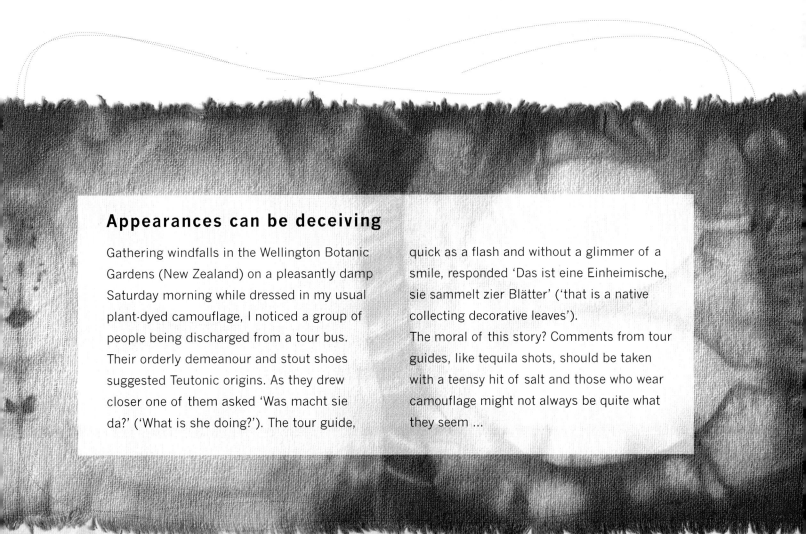

Appearances can be deceiving

Gathering windfalls in the Wellington Botanic Gardens (New Zealand) on a pleasantly damp Saturday morning while dressed in my usual plant-dyed camouflage, I noticed a group of people being discharged from a tour bus. Their orderly demeanour and stout shoes suggested Teutonic origins. As they drew closer one of them asked 'Was macht sie da?' ('What is she doing?'). The tour guide, quick as a flash and without a glimmer of a smile, responded 'Das ist eine Einheimische, sie sammelt zier Blätter' ('that is a native collecting decorative leaves').

The moral of this story? Comments from tour guides, like tequila shots, should be taken with a teensy hit of salt and those who wear camouflage might not always be quite what they seem …

As a small girl I loved to sit on the front step with my grandmother, listening to her telling stories as the sun gradually set. My favourite was that of the princess who finds herself alone in the forest and must make her clothing from what she can find – leaves, grass and wisps of fur caught on the bushes. I imagined the dress of leaves pinned together with thorns, bejewelled with luminous beetles and dewdrops. This was my dream dress and featured frequently in the abundant drawings that covered the pages of my schoolbooks.

Although this is merely an old story, it does bear some relation to fact. Until the development of chemically synthesised fabrics in the twentieth century, cloth was indeed made from substances found in nature. Essentially animal hair, animal skins, plants and seeds provided the fibres from which clothing was made. Fish scales, carefully cleaned, did service as sequins. Shell and bone were carved into buttons. I learned much later in life that colourful beetle wings were indeed used to embellish clothing and jewellery in Mogul India; they were imported by the millions from Burma where the adult beetles, having fulfilled their function in life by mating and laying their eggs, conveniently expired and could then simply be swept up and bagged for export. The magnificence of such Indian textile works was not lost on the British; jewel-beetled fans, ornaments and tea-cosies enjoyed considerable popularity in Victorian England.

jewel beetles such as this one were also exported from India to Britain in their hundreds of thousands

garment of weeds stitched with linen thread ▶

chapter 1 Underpinnings

Another faerie tale – that of the sister condemned firstly to spin then to knit twelve garments from nettles for her twelve brothers – also has its roots in reality. Nettle fibre has been used to weave cloth throughout history and is presently enjoying something of a revival in Europe. The laborious gathering of nettles, regarded as weeds of wastelands and waysides, had its eventual reward (after a lot of hard work) in soft, beautiful and resilient cloth that lasted for years.

The origins of textiles

While it is generally assumed that fabrics began with the making of yarn and that flax is the oldest known fibre, my hypothesis is that felt (the friction-induced meshing of wool or hair under warm and damp conditions) is likely to have been the first constructed textile used by humans. Imagine the cold, dank confines of cave life back in the glimmering dawn of time. Curling up by the fire sharing a morsel of mammoth with the recently tamed wolf, our prehistoric human might well have enjoyed the shed fur that is an inevitable consequence of keeping animals in a dwelling place. The fluff may have been gathered together to make a softer place for bodies to sleep at night, and given the sorts of activities undertaken by humans in keeping warm, amusing themselves and ensuring the proliferation of the species, our early homemakers may have accidentally discovered the joys of the fused textile that we know today as felt.

a richly stitched fragment from a central Asian garment (left, front; right, back), early twentieth century

Sadly, though, because felt is formed of animal protein, it is also a delicious source of nourishment for micro-fauna (moths, grubs and bacteria) as well as a splendid nitrogen-rich slow-release fertiliser for plants. This would explain its extreme susceptibility under archaeological conditions and hence the very limited examples that survive today. Those pieces that have survived have done so because they were preserved in permafrost (for example, the Pazyryk felts in Russia) or in highly saline and very dry soils (the mummies of Ürümqi, found in China).

Our omnivorous forebears also realised that the fur encasing their meals-on-the-hoof could be treated to make it last longer and could be wrapped around their own more delicate covering as protection against the elements. No truly accurate dating of textile-related developments is possible, as the objects themselves were relatively fragile compared to the durability of wood, metal and stone. The clues that remain for us are the tools that might have been used and remnants such as the impression of a piece of twined string in a shard of pottery. One of these has been estimated to be about 20,000 years old; however, it is not inconceivable that the idle twining of fibres that led to the discovery of string well predates the firing of clay as an artisanal activity. Museum exhibits of early pots show quite a few that appear to have been formed of raw clay built around a basketry framework.

So at some point a very long time ago a person twiddling idly with bits of grass, hair or vine realised that the more filaments that were plied together, the stronger the resulting cord would be. Textile scholar and weaver Elizabeth Wayland Barber puts it this way: 'somewhere between twenty to thirty thousand years ago … some genius hit upon the principle of twisting handfuls of little weak fibres together into long, strong thread'.[7] I am inclined to agree with her that string ranks as one of humankind's most significant inventions. What might have seemed at first a relatively minor development meant that objects could be tied together, weapons and tools made, animals (and possibly other humans) restrained, nets and fences constructed, individual animal skins joined to make larger protective sheets and eventually cloth woven with the development of the loom. This cloth could be formed into shelter as well as clothing. The rest, as they say, is history;

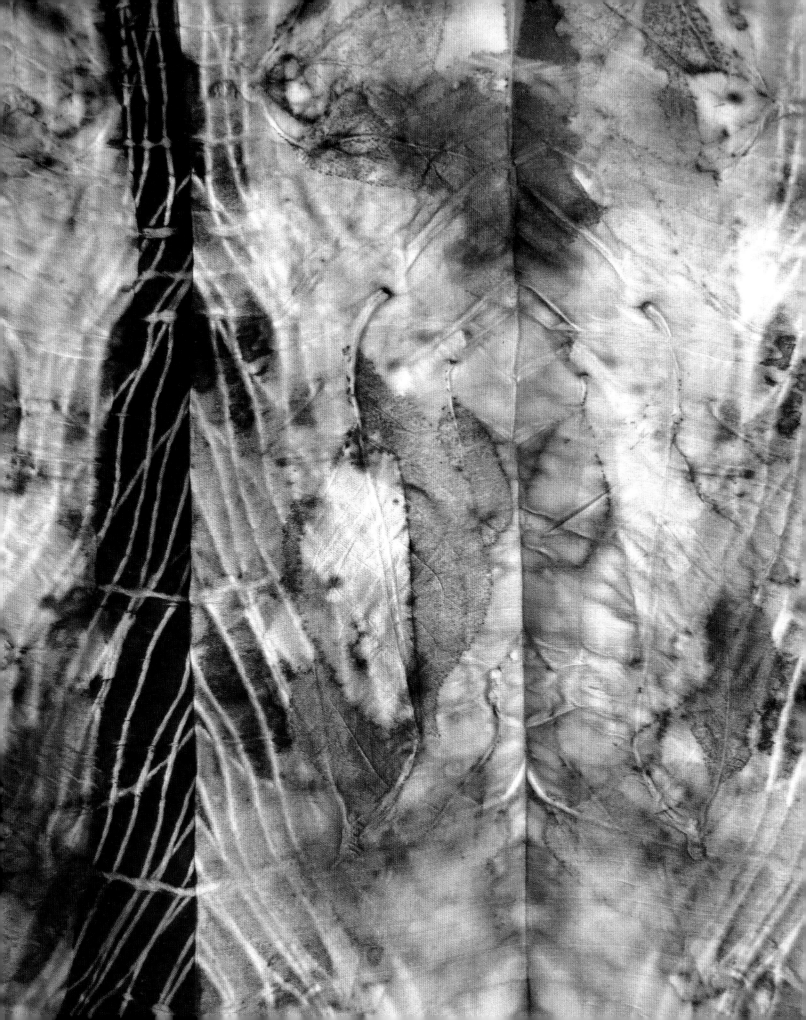

one so important to daily human life that there are many words and phrases used in conversation that have been derived from things to do with cloth and fibre. Think about such terms as 'spinning a yarn', 'having the wool pulled over one's eyes', 'wool-gathering', 'losing the thread', 'knitting together', 'airing one's laundry' and so forth.

It is, moreover, a history that – although relatively brief in relation to the age of the planet – comprises inventions, developments and practices that have brought about radical changes in our environment and may well put the future of our children at risk.

Of course the textile industry is not the sole culprit in regard to the misfortunes of the world, but it does play a leading role in the unfolding drama. By making do with having a bit less as well as making that 'less' last longer, each person can do something useful, simply by moderating their impact on the planet. In this book I'd like to take the reader on a journey that encompasses the growing, harvesting and making of fibre and cloth as well as maintenance and mending, embellishment, repurposing and dyeing. I think that bringing back grandmother's philosophies – of valuing beauty, practising thrift, being mindful, consuming and making less but doing it more slowly and better, being content with enough – can not only help to make the world a better place, but are principles with the potential to enrich our lives. I'm a very long way from doing everything properly but I like to think that I'm giving it my best shot. There are those who will argue that publishing illustrated books such as this one is a huge waste of resources and in a way they have a point. On the other hand, if this book saves you some money, helps you extend the life of your clothes and brings you delight in the dye-pot, then I shall have achieved something worthwhile.

I have seen a many wonders, and have a good memory for them; and in spite of all grumblings have a hope that civilized people will grow weary of their worst follies and try to live a less muddled and unreasonable life. [8]

◂ windfall cloth, San Francisco

There are so many things that we do simply because we feel obliged to or have been burdened with guilt. We can bicycle or recycle out of a sense of fault or fear for the present state of the world or we can make the most of life and go through our days mindfully, performing the various tasks before us with joy.

The poet and writer Robert Louis Stevenson said of travelling that he travelled not to go somewhere specifically but simply to go; that it was about the act of the journey. It is also a useful metaphor for life and for making. When we make things such as our clothes or objects for our homes with care and consideration we are able to experience the contemplative pleasures of making things with skill by hand. I sometimes wonder whether as a society we are not only often too removed from nature because of the ways in which we live but if sometimes we even suffer from a kind of tactile deficiency. The mindful making and doing of things, even menial ones, has the potential to enrich our lives on many levels.

When we adopt this approach, even the process of washing dishes becomes an act of gratitude, as the very reason we are faced with the task is that we have been able to enjoy a meal. I'd rather eat and wash dishes as a consequence than go hungry! Similarly the patching and mending of clothes need not be a mundane occupation. The *boro* textiles of Japan and the *kantha* cloths of India, Pakistan and Bangladesh show so beautifully how almost inconsequential scraps become exquisite narrative-rich objects when pieced, layered and reinforced with stitch.

◀ one of several worktables in my studio

chapter 2
Provenance

The web of our life is of a mingled yarn, good and ill together.

William Shakespeare, *All's Well That Ends Well*, Act IV, Scene III

my tireless sheepdog Kip, shadowing her woolly companions

In this chapter we will take a look at what textiles are made from and how they are put together as well as investigating some of the provenance of dyes and trims. Garments might be made from ethically and ecologically sound fabrics but the threads with which they are constructed as well as the trims that are applied often have other stories to tell.

Looking at the contemporary fashion world, we find the descriptor 'organic' applied liberally, largely without due consideration as to what this term actually means. In essence, if something is 'organic' then it either is a living organism or was originally derived from one. The field of organic chemistry was once defined as the chemistry of substances produced by living organisms, but in contemporary parlance is more accurately described as the chemistry of carbon and now extends to include substances that have been synthesised artificially. Under the latter qualification synthetic fabrics derived from petrochemical compounds could in theory be perfectly logically classified as organic; oil, after all, is a substance formed from matter that was originally living. In the modern vernacular, however, 'organic' has come to mean something that is 'clean, green and pure', untainted by chemicals and in theory 'ecologically friendly'. This is where we begin to slip slowly into the swamp. Even the use of the word 'chemicals' might be questioned. After all, everything in the world, humans included, is made up of chemical compounds of varying complexity, mostly existing in a kind of symbiosis – getting on with business side by side with other bits also made up of elemental compounds. The air we breathe is a cocktail of various gases. Pure water is a compound of two elements, oxygen and hydrogen, which are also found in gaseous form in air. But to keep things simple let us assume we can accept the appellation 'organic' for a product that has been produced without the intervention of synthesised chemicals and take a closer look at some of these things.

Organically grown cotton, for example, might have been grown without the application of herbicides, pesticides and synthetic fertilisers; but what happens when it leaves the farmer's field? It goes to be processed (usually in some kind of oil-consuming mechanised transport) in a mill that also consumes some sort of artificially generated power. The fibres themselves need softening to make them acceptable for our delicate skins; these are processes that also use chemicals (as well as more energy). What about the finishing of the shirt? There may be the application of a dye to satisfy the consumer's passion for colour, as well as buttons, stiffening, packaging and labelling to distinguish the product from the competition, and still more transport involved to get the goods to market.

The supply chain that delivers the garment to the store where you can buy it is fundamental to the story. Tracing the provenance of your consumables isn't always easy and sometimes seems almost impossible; in the end it is up to individual consumers to satisfy themselves as to whether they wish to endorse a particular product or brand by acquiring it and wearing it.

Natural fibres

Natural fibres fall into two broad categories: those derived from organic (living) sources and those made from inorganic substances (such as minerals or metals). The former category can be subdivided into fibres produced from animal proteins and those culled from the plant world, which can be split again into a number of groups: those sourced from the seed of the plant, and fibres from stems, fruits, leaves and even the sap. Rubber, although feeling quite synthetic, is a natural fibre produced from the sap of the rubber tree.

fabric from beans? Yes indeed, but no simple process, nor without ecological consequences

Not all natural fibres are necessarily sourced by ecologically sustainable means or indeed even safe to wear. Asbestos, for example, is not a substance that one would choose to wear. Soy silk, although sounding as though it could be a delightful cloth made without compromising the wellbeing of the silk moth, can undoubtedly have a deleterious effect on the environment as forests are clear-felled so that soybeans can be cultivated in place of trees.

The same cautions apply to cloth made from maize, especially as the cultivation of corn very quickly depletes soils of nutrients, leaving the land unproductive and thus encouraging the compensating application of fertilisers with all their attendant effects on waterways.

The following table outlines at a glance the various derivations of fibres.

Natural fibres	Organic derivation [9]	Vegetable	**From seeds** Soy silk
			From seed capsule Cotton, kapok
			From bast or stem Flax/linen, hemp, jute, ramie, kenaf, straw, banana, pineapple, papyrus, esparto, sugar cane, wisteria, nettle
			From leaves Aloe, yucca, elephant grass, harakeke
			From fruit Coir (coconut)
			From plant juice Natural rubber
			Various Peat
			Bark Paper yarn
		Animal	**Clipped or combed from the living animal** Sheep wool, camel hair, vicuña hair, alpaca hair, llama hair, goat hair (mohair, cashmere, angora), rabbit hair
			Shaved from the deceased animal Possum fur, rabbit hair
			Insect Silk
			Milk Casein thread
	Inorganic derivation	Mineral	**Mined from the earth** Asbestos
			From metal Metallic

Linen

Linen cloth is derived from the flax plant (*Linum usitatissimum*) and has been used by humankind since the Palaeolithic era. Excess seeds of this useful annual were once used in bread baking or boiled into a jelly and given to stock (apparently they are poisonous to horses if eaten raw). The extraordinary Hildegard von Bingen (1098–1179) – composer, writer, ethnobotanist and founder of convents, who also discovered that the inclusion of hops in the brewing helped to preserve beer – allegedly used flax meal in hot compresses to treat both internal and external ailments. While her music was exquisite, some of her alleged pronouncements – such as (in regard to what conditions might determine a child's disposition), 'the worst case, where the seed is weak and parents feel no love, leads to a bitter daughter' – could be taken with a teaspoon of salt. The hot linseed compress, on the other hand, may have been the mediaeval version of the popular heated wheat-bag.

My great-grandmother, who lived on a small farm in Latvia, cultivated this plant as part of the farm garden along with the vegetables and cereals necessary to support the family and animals. Each year the plant was harvested, retted (allowed to partially decompose) in a pond and the fibres then rolled before being drawn through a wooden hackle. She refined them by hand to an astounding degree of smoothness, weaving sheets and making fine embroidered shirts and blouses. I treasured for years a nightshirt that had been sewn from sheets woven by her and passed to me by my mother, who had worn it after my aunt, for whom it had been made as a gift for her sixteenth birthday by my grandmother … who had packed the 'second-best' sheets when they fled Latvia (the sheets from which the nightgown had been cut had already been turned 'sides-to-middle'). It was a deliciously comfortable garment but although the intermediate materials had been carried safely out of Latvia during World War II and on to Australia in reconstructed form without damage, it sadly did not survive the Ash Wednesday bushfires of 1983.

Linen is a wonderfully strong fibre that can be spun to a fine but durable thread. Waxed linen was traditionally the thread of choice for attaching coat buttons and sewing leather and canvas goods as well as human flesh (in the form of surgical thread). Linen cloth is available in the finest of handkerchief muslins or, at the other end of the spectrum, as sturdy furnishing fabric suitable for upholstery. The famous Sopwith Camel biplane flown by British pilots in World War I consisted of a wooden frame covered in Irish linen (sealed with nitrate dope).

The flax is harvested by pulling the whole plant (an annual) from the ground and the processing begun by simply allowing it to ret (whereby the cellulose material rots away, revealing the sturdy fibre) in damp fields or in a purpose-dug pit. In Egypt the stalks are still retted by long soaking in the River Nile. This might not necessarily be an ecologically sustainable practice, as introducing biological nutrients into a waterway could support the growth of algae – but on the other hand, chemical processing would probably have far greater impact through polluted emissions.

Harakeke

Phormium tenax, also known as harakeke or New Zealand flax, is actually a kind of lily with a very fibrous long leaf that is suitable for weaving when split and also for making strings and yarns when scraped clean and stripped to its core. This versatile plant is a source of medicine as well as fibre for shelter and clothing, provides food for the tui (a beautiful blue-black bird) and yields rich colour for cloth. It has also been extensively used for papermaking.

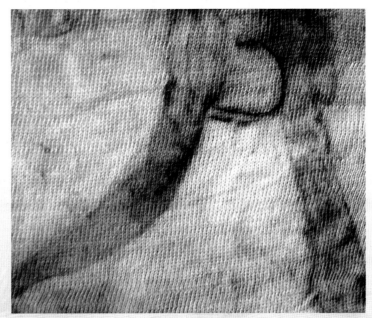

a print on merino jersey from the spent harakeke flower (the blue imprint is from the rind of the stalk)

Hemp

Hemp (*Cannabis sativa*) is another very valuable annual plant that once enjoyed popularity in many parts of the world. It is only comparatively recently that it has fallen into public disfavour (or rather been proscribed by those in power) due to the psychotropic properties of specially bred strains. It is believed to have originated in the Himalaya, and its local name, *ganja*, comes from Sanskrit. In rural Europe it provided useful fibre, mostly employed in the making of sacking and ropes. Christopher Columbus's sails and ropes were made of hemp.

Surplus green material from the fast-growing crop can be usefully applied as mulch, and oil from the seeds is rich in things that are good for the body. In Latvia the seeds are still used to make a dark and delicious paste not unlike peanut butter (but be careful not to carry it in your luggage when in transit via Singapore) and traditionally older people suffering from rheumatism smoked the (only very mildly psychotropic) leaves in their pipes as a form of gentle pain relief. It was a generally handy crop and in temperate climates rarely required extra water during the growing season. It seems quite ridiculous that hemp seed can be processed for pet foods in Tasmania but is banned from inclusion in comestibles destined for humans.

In Japan hemp was also a common agricultural plant, especially in the cold north where it was impossible to cultivate cotton. Then in 1948 the United States, as occupiers of the conquered country, rewrote the Japanese Constitution and published the *Hemp Act*, under which cultivation of this important species was banned. With the stroke of a pen it became forbidden to grow hemp, virtually wiping out a source of paper, cloth, rope, stockfeed and birdseed. Fortunately the species can still be sourced from wastelands and rail sidings, as otherwise this plant (so perfectly acclimatised to its bio-region) could have been lost completely through General MacArthur's ignorance.

naturally dyed hemp jersey ▶

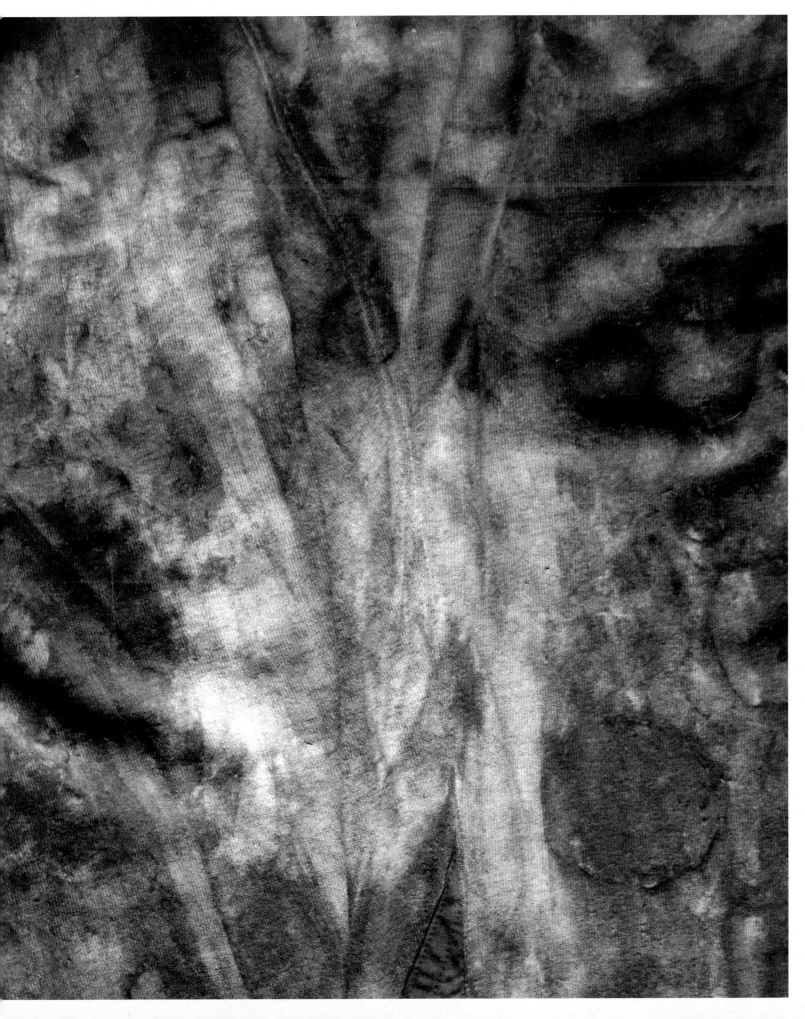

Cotton

Cotton (*Gossypium* sp.) is a rather thirsty plant that was apparently first used for cloth making in India, whence it spread to the Middle East and China. The fibre used is taken from the cotton boll, a fluffy growth harbouring the seeds of the plant. These fibres are quite short and fluffy, can be challenging to spin by hand and (unlike those of hemp or flax) can cause severe lung problems for those working with them. The genus *Gossypium* is a member of the Malvaceae family, along with species including hollyhocks, cacao, hibiscus, marshmallow and okra.

wool and cotton blends were popular in the making of sturdy insulating undergarments

Asian species include *G. arboreum* and *G. herbaceum*. The genus became popular in ancient Egypt, flourishing along the fertile banks of the River Nile where the soils were annually replenished by rich silt deposits from flooding. Museum examples embroidered with indigo- and madder-dyed threads reveal beautifully fine cloth woven with great dexterity. Egyptian cotton sheets are still much prized in the bedroom, but these days are made from selectively bred longer-fibred species introduced from South America. Australia itself has some seventeen endemic species while Mexico and Northern Africa boast similar numbers, although it is believed that some coastal populations in the former may have established themselves by floating in on the tides from Central America. Maybe they were long-distance trans-oceanic immigrants like the coconuts that sprout on foreign shores after floating the seven seas.

clockwise from top: cotton being packed in an Indian village; a length of cotton in the form of a sari; village sheep in India ▶

South America is home to the species *Gossypium barbadense*, known for its long beardlike fibres (as its name so graphically indicates) and endemic to Peru and southern Ecuador. This variety and *Gossypium hirsutum* constitute the main commercial cotton crops today. With the discovery of the Americas and the establishment of cotton cultivation, various unscrupulous and enterprising persons also established the slave trade, capturing and delivering Africans into lifetimes of abject misery and deprivation as free labour for rich plantation owners.[10]

Cotton was introduced to Australia in the 1860s when the American Civil War interrupted supplies to British mills, but it was not successfully established in the country as a recognised industry until a century later. It is one of the most damaging crops grown on the driest continent, absorbing vast quantities of water and requiring repeated applications of toxic chemicals to keep insects at bay. Defoliants are applied at harvest time to ease the process of mechanical picking. Such chemicals inevitably have an impact on the land, accumulating in the soil and in the wild fauna inhabiting the area. Some years ago beef cattle being fed cotton trash to supplement their diet during a prolonged drought were found to be unfit for human consumption due to high levels of a chemical known as Helix (chlorfluazuron) that had originally been applied to cotton crops through aerial spraying. Cows consuming this feed also passed the toxins to their suckling calves. The drift from such aerial spraying must also be cause for serious concern. Industry websites suggested that the contaminating residues might never be metabolised by those cattle and consequently endorsed their slaughter for pet food. When we consider the longevity of this toxin and how it might then transfer through the pet and be distributed as faecal matter as well as (later) the corpse of the ultimately deceased pet, one begins to appreciate how such things can gradually pervade all facets of life. Bear in mind, too, that poisons applied to the plant will probably also be present in the fibre derived from that plant.

On the agricultural scale, such as when chemically tainted feed was supplied to cattle, not only were the cattle themselves affected, their emissions eventually tainted nearby waterways as well as the aquatic inhabitants. Australia may have banned that particular substance, but experience has shown that this may simply have shifted the problem elsewhere. Who knows what is being tipped onto fields in such countries as India and China? The history of the draining and destruction of the Aral Sea is testament to the impact that unrestrained agricultural practices can have on the environment.

'Organic' cotton (that grown without the application of pesticides, herbicides and synthesised fertilisers) requires the same volume of irrigation water as that grown with chemical assistance. Once harvested it undergoes similar processes to its rival, which may include the application of colour using synthetic dyes usually derived from petrochemical products. Such colours are often described as 'safe' by the industry, but the effluents from dye houses are still a source of pollution. Until consumers are assured of supply-chain transparency it certainly might feel better by buying organic, but it may not have as much positive effect on the environment as we would wish. Nonetheless, purchasing 'organic' cotton remains very much the lesser of the two evils provided the cloth has not been dyed using synthetic colourants.

Gradually a range of appellations is beginning to emerge, including 'fair-trade cotton', 'organic cotton', 'low-water cotton' and 'low-chemical cotton'.

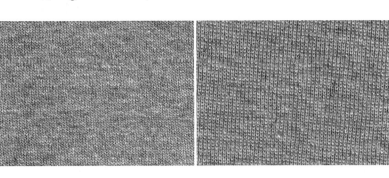

natural colour organic cotton – 'grown green' and 'grown caramel'

Banana

Banana (*Musa* sp.) fibre is a useful by-product of fruit growing. The long stalks of the banana plant (often mistakenly referred to as a tree) contain bast fibres of great beauty that can be used to weave a shiny cloth popular in some Asian countries for celebration garments, although it is more commonly used to make paper and products as diverse as teabags and shoes. The traditional Japanese textile *bashofu* is woven from banana fibre. It makes a good deal of sense to use the whole banana plant, especially as it is lopped to the ground anyway in the banana harvesting process.

Ramie

Ramie (*Boehmeria nivea*) is a hardy rhizomaceous perennial that is well suited to tropical climates and can be harvested up to six times a year. Although one doesn't generally think of Egypt as a tropical country, this crop was apparently grown there between 5000 and 3500 BC and used for making mummy cloths. Modern processing involves chemical treatment, as degumming is required, but the fibre is naturally pure white in colour, which would suggest that bleaching is unnecessary. The fibre is quite long, durable and silky as well as being extremely absorbent; this latter property explains why it takes dyes so well. It has good resistance to insect and bacterial attack, increases in strength when wet and becomes more comfortable to wear as it ages. Ramie has a tendency to crumple easily but responds well to cool washing and line drying. The filament is quite often blended with cotton as it gives better strength to that relatively weak fibre.

Nettle

Nettle (*Urtica dioica*) fibre, while regarded as the stuff of faerie tales, is a very real alternative to crops such as cotton. Nettle cloth was apparently

some say nettles are weeds; others are grateful for their many uses such as in brewing tea, as a cooked green vegetable, as a source of green dye and as a resilient fibre for making cloth much used by the German armed forces in World War II as it became difficult to obtain cotton for their uniforms; the control of the supply of the latter was largely in the hands of the British.

Fabric producer Brennels (in the Netherlands) is working on the development of anaerobic digestion processes in order to produce light and delicate fibres, monitoring the progression of the breakdown to determine the point at which the fibres would have optimum handle while still retaining their strength. Stopping the system at the right point is critical to having the most beautiful fibre. Although the company has no pretension toward organic farming, they are looking for an alternative to cotton production, stating that nettles require less water and chemicals during cultivation anyway. The good news is that the anaerobic digestion system makes use of natural enzymes that are already part of the nettle plant, thus saving on adjunct chemicals.

Bamboo

While bamboo (*Bambusa* sp) is essentially a fast-growing genus of giant grasses and very much a renewable resource, large areas of rainforest are being felled and planted with bamboo to feed the rising demand for fibre production. Destroying viable ecosystems in order to superimpose monoculture crops certainly isn't ecologically sustainable.

In Asia the more substantial stems are used for building houses, fences and quite spectacular scaffolding. Young fresh shoots can be eaten (and not just by pandas). Fabrics made from this plant group are described as 'green and sustainable', but there's a catch. Certainly the cloth that is produced from the actual fibre of the plant can be described as a sustainable product. However the soft, easily draped variety of bamboo cloth is produced in a similar way to rayon (a synthetic fibre derived from wood) whereby the plant is processed using a range of toxic chemical solvents (with all the attendant consequences) together with considerable heat energy. The resulting fibre bears no resemblance at all to the original material, having been extracted into liquid form and then spun. This hardly seems to fit the description of a 'natural fibre'. The difference will be in the fine print on the label – 'bamboo fibre' is quite a different product from 'fibre derived from bamboo'.

Not greed nor gold nor gadgetry, nor principalities nor powers, including all the combined resources of the entire artificial fibre industry have been able to synthesize a single solitary strand of wool. [11]

once scoured, or washed, this merino fleece will be pure white. Wool picks up dust from the country the sheep graze and in Australia it is possible to guess at the provenance of sheep depending on the hue of their wool, which can be anything from pinkish red to dark grey. The colour is usually only on the surface (the tips of the wool) and when the fleece is parted the wool will show creamy white

Wool

I must admit to being slightly biased, as despite being sensitive to prickly fabrics I am firmly convinced that wool is a wonder fibre and also that far from being stupid the sheep is actually an intelligent and amusing animal. Some of them even have a sense of humour. We had a ram once who would stand in the sheep yards, cheekily open a gate latch with his teeth and then not bother escaping but simply stand there gazing limpidly into the distance. Another sheep that had been rescued as an orphan and raised on a bottle had a habit of shadowing me whenever her mob was in the yards. Whatever I did, Fuzzy was 'in behind' just like a good border collie dog, except rather larger. British-bred sheep seem to have a particular affinity for human company and prefer being led to being driven. When I have the time and the weather is nice I like to take a book or journal out on the paddock and find a comfortable spot to sit and read or work. Inevitably the flock will gather around and settle down just like a friendly collection of companionable clouds. It is particularly entertaining at lambing time as the lambs romp and churn about in between the ewes, leaping on and off the boulders (and sometimes even their snoozing mothers) in woolly waves.

◀ charming English Leicester sheep at Hope Springs

Environmentalists often deride wool, as the sheep that produce it have gained a reputation for destroying country with their hard feet and hungry mouths. Those managing the flock should be held responsible rather than the creatures themselves, who clearly have no control over their destinies. Certainly when farming country is cleared, over-grazed and exploited beyond its carrying capacity disastrous consequences can be expected.

But if pasture is managed properly and includes a light tree cover to provide shade and help stabilise the soil then it could be argued that sheep are efficient processors of carbon (in the form of grass) into protein (in the form of wool and meat). Moreover the trees will act as natural air-conditioning systems to help keep ambient temperatures lower; also, they may well help generate precipitation and will certainly prove valuable as long-term carbon storage.

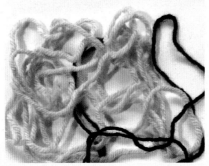

organic wool knitting yarns in naturally grown colours

Wool is truly a miracle fibre. It accepts plant dyes very well, especially those sourced from eucalyptus. Being a hollow fibre it is an excellent insulator and, when not processed with chemicals to attain 'superwash' status, is also fire resistant. Untreated wool must reach temperatures of at least 570°C (1058°F) before it will ignite; cotton will burn at only 255°C (491°F). Wool then has another trick up its sleeve; as it burns, the fibre releases a nitrogen-rich foam that actually helps to smother the flames by blocking access to oxygen. This is why wool will smoulder rather than actively burn. The biggest advantage to wearing wool in situations of fire danger is that, unlike synthetic fabrics, pure wool will not melt and so will not stick to skin. The temperatures required to actually destroy wool would mean certain death for fragile living organisms such as us humans. It is astonishing that the fire-fighting services in Australia have not adopted wool, the miracle self-extinguishing fibre, for use in their protective uniforms.

for wool to be machine washable it must be treated with organic chlorine compounds to wear away some of the outer layer of the fibre (removing the tips of the scales) and make it less likely to felt. Unfortunately this also affects the potential fire resistance of the fibre as well as making the wool feel harsher to the touch

The thickness of wool is measured in microns (thousandths of a millimetre) and ranges from superfine merino at 13 microns to the strong lustre wools of the Lincoln at 44 microns. The merino sheep produces wool that is ideal for clothing such as undergarments, wraps and soft T-shirts as well as coats and jumpers. Wool grown on the Lincoln sheep, at the other end of the spectrum, has for years been used to make wigs for both people and dolls. It also makes splendid long-lasting furnishing fabrics.

In Australian wool-classing terminology, any breed that is not a merino is classified as a cross-bred, irrespective of their pedigree. The amusing thing here is that most of the British breeds have lineages that can be traced back over hundreds of years, whereas the large-framed merino so popular in Australia owes its stature and wool-producing ability to early cross-breeding with large British breeds such as the English Leicester and the Lincoln. There's a memorable passage in one of the 'Billabong' series of children's books by Mary Grant Bruce in which the Linton family are returning to Billabong Station in 1919, having patriotically served King and country by supporting the war effort in Britain. They are described travelling through Victoria by train and commenting on the small stature of the merino compared to sturdy British sheep they had become accustomed to: 'pygmies, after the great English sheep'.[12]

Wool is an incredibly versatile and useful fibre. It can be spun, plied, woven, felted, knitted or crocheted. It can hold up to 30 per cent of its own weight in moisture before it begins to feel damp. When it does become damp, however, it will still feel warm, due to a complex chemical reaction between water and the molecules of the wool. The nature of the fibre is such that the myriad tiny air pockets trapped within it not only act as insulators but also have a wicking effect, drawing moisture away from the skin and into the fibre. The fibre is also remarkably elastic, which together with its other properties makes it a brilliant fabric for both active wear and everyday wear.

The UK company Second Nature[13] is utilising the thermal properties of wool in the production of compressed wool insulation for buildings. Unlike fibreglass insulation, which often causes respiratory problems, this product can be handled without the need for personal protective equipment.

well-insulated Swaledale sheep in the snow

Looking closely at the individual wool filaments one can see that they have tiny scales (rather as fish do). These are important in helping spun wool to retain its twist and are the reason felt can be made from wool. Felting requires friction, moisture and wool, with heat as an optional extra (it isn't essential but it helps speed up the process). As friction is applied in the form of jiggling, throwing, dropping or rolling, the fibres begin to move, but just like fish they can only wriggle in one direction. This is why wool appears to shrink during the felting process. It is also one of the reasons it is so difficult to restore a shrunken woollen garment to its former size. Another factor is that when wool becomes wet the small scales swell and lift. When it is dried again the scales lock down over any fibre that happens to cross them. Wool thus has a 'memory' for process. Those little scales also help to lock in dye colour if the wool is treated correctly during the colouring procedure.

While processing wool does indeed require large quantities of water and energy as well as presenting the challenge of disposing of the lanolin-rich effluent, the material itself embraces so many valuable qualities that I think on balance it can be justified as a sustainable product. Providing no toxic dyes and finishes are applied, a woollen garment can be mended, unravelled, re-knitted, felted or woven, patched over and over again and eventually applied to the garden as a nitrogen-rich, slow-release-fertilising weed mat.

a Chocolate Wool Gotland sheep flock near Palmerston, New Zealand

A myth that has been gaining in popularity is that shearing sheep is cruel and that the wearing of wool should be avoided. The truth of the matter is that humankind has been selectively breeding sheep for some four thousand years, developing them from creatures that initially had a coat much more akin to that of a dog (with sturdy guard fibres over a soft undercoat) to the soft and long-woolled animals we know today. If we don't shear their wool annually they will die gruesomely through flystrike, become wool-blind (the wool grows over their eyes so they can't see) or simply become so heavy that they can no longer get to their feet and walk about (especially when wet). Some years ago an elusive merino wether became famous in New Zealand when it was found after eight years unshorn in the wild mountains of the South Island. The animal looked like a wandering volcanic ash cloud and had a drastically reduced field of vision thanks to its thick fleece.

just a few of the many naturally coloured shades of wool

Anyone who has ever had their long hair cut short at the hairdressers will be able to appreciate the lightness and elation this animal must have felt when finally relieved of its burden.

Officially Australia has a relatively small 'organic' flock; however, the sheep being run in the pastoral country of the interior are rarely (if ever) dosed with chemicals. They also have ample space to roam and so avoid infestation by worms and other parasites as they do not graze areas that are contaminated by faeces. The wool cut from these sheep is, chemically speaking, extremely clean and conforms very closely to standards we would regard as 'organic'.

Given the high ecological cost of dyeing wool black for suits, it would make sense to expand production of naturally grown fine coloured wool. Traditionally woolgrowers have slaughtered coloured sheep as soon as they are of edible size to avoid coloured fibres being introduced into their clip. Breeding sheep for coloured wool beyond the hand-spinner market would obviate or at least reduce the need for toxic dark dyes.

Mohair, cashmere, angora

The soft, lustrous fibres produced by various breeds of goat share many of the eco-sustainability properties of sheep wool. Their generally finer epidermal scales make them appear quite shiny and silky (although the cut ends from shearing may still tickle!). Goat hair has been a staple fibre of the Middle East for thousands of years and even gets a mention in the Old Testament.[14]

It is believed that the name mohair is derived from the Arabic *mukhayar* or *makhayar*, possible translations of which include 'selected choice', 'best of selected fleece' and 'cloth of bright goat hair'.[15] Unfortunately, as well as these alluring and creative descriptors for their hair, goats also have a well-deserved reputation for eating anything and everything, so it is not entirely out of the question that this genus might well have played a substantial role in the desertification of the once-fertile crescent of lands embracing the rivers Tigris and Euphrates. Artificially irrigated agriculture must also carry a share of the blame for the state of the region, popularly supposed to have been the birthplace of civilisation.

Goats, however, are generally quite hardy animals, often with delightful personalities. Watching herdsmen and shepherds passing each other on roads in India is fascinating as the belled flocks filter happily through each other and the animals make sure not to lose sight of their assigned guardian.

Alpaca

Alpaca, llama, vicuña and guanaco are all soft-footed camelid species that have been husbanded in South America for thousands of years. The fibre varies from the quite hairy guard hairs of the llama to the silky softness of the suri alpaca. Alpaca is even more thermally stable than wool, apparently water repellent, very light in weight and considered virtually hypo-allergenic as it does not contain lanolin. On the other hand I have been assured that it is difficult to spin and often quite filthy as the animals like to loll about in dust-baths. Garments made exclusively from alpaca tend to stretch (particularly when wet) and felt made only with alpaca can be pulled apart quite easily as the fibre is so smooth.

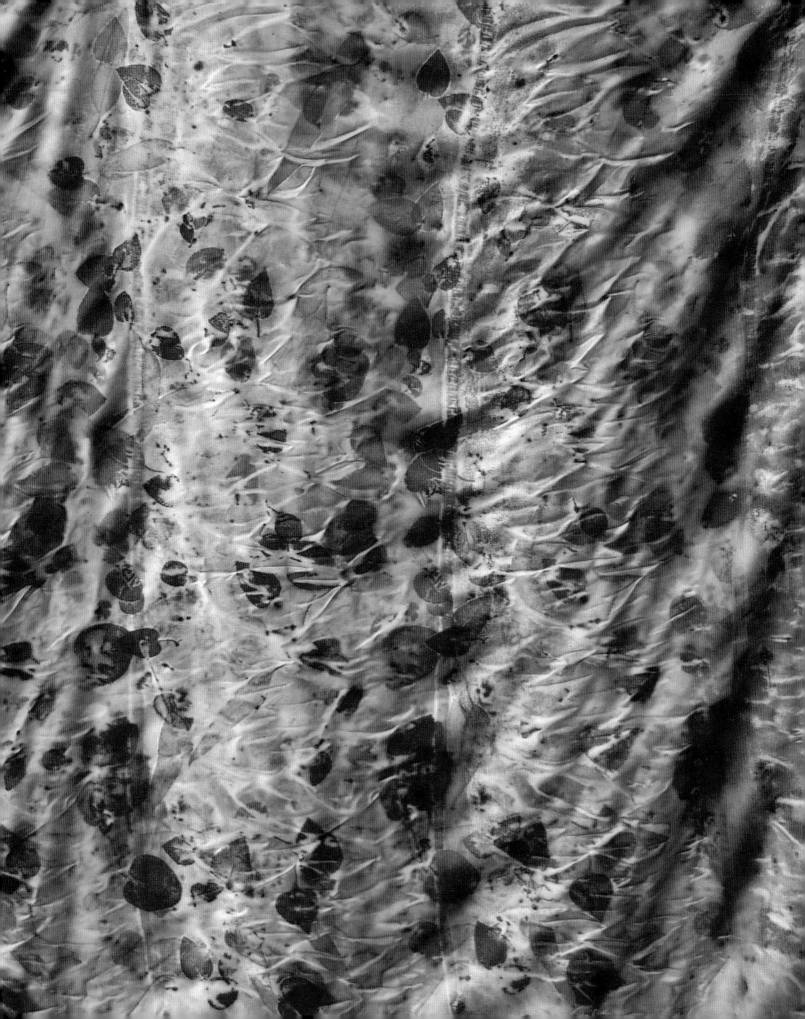

soft China silks resting on a piece of hand-woven Japanese hemp

Silk

Silk is a protein substance very similar to human hair. It is exuded in liquid form, hardening on contact with air, by the caterpillar of *Bombyx mori* in the construction of the cocoon in which it pupates before emerging as a blind, flightless moth whose sole purpose during its very brief life is to mate and reproduce; thousands of years of cultivation and continuous hand-feeding have led to the development of an animal utterly dependent on human intervention. (Those humans hunched over their laptops and looking up only for cups of tea and biscuits could well take the silk moth as a cautionary example.) The eggs of the moth are tiny; under the right conditions a mere handful will develop into tens of thousands of hungry caterpillars requiring the regular delivery of tonnes of chopped fresh mulberry leaves. Silk is one of the least polluting fibres, requiring in its most basic form only boiling water for the degumming process, although I must admit to some serious misgivings about the condition of the hands of those who commence the unreeling. In China sericulture has been practised for thousands of years, primarily the work of women who would spend the better part of their year feeding caterpillars or unreeling, spinning and weaving silk.

◀ natural eucalyptus prints on silk cloth

silk cocoons photographed in India

Strands from several cocoons are reeled together to form a thread that is interwoven in various ways to create the structure of cloth. The tiny prisms that comprise the structure of the silk fibres generate the sparkle and shine on the surface of the fabric. The rougher the silk cloth, the more intensely it seems to take dyes; silk noil fabric and silk organza tend to dye much more vigorously than silk satin (on the latter the brightest colours are always on the reverse, or rougher, side of the cloth). China managed to keep the mysteries of sericulture to itself for a long time but the secret (and the necessary silkworm eggs) eventually spread, first to Korea and then Japan and India. It is not surprising that silk production has nearly doubled in the past thirty or so years as silk is a delightful fibre to wear. Like wool, but for different reasons, it feels warm in winter and cool in summer. Babies wrapped in silk quilts seem to fall asleep instantly, I think because the touch of silk on their cheeks feels much like that of human skin. My three children each had silk quilts in their infancy and each wore them out several times, they loved them so much. A filament of silk might be stronger in tension than the same diameter filament of steel but it takes more than tensile strength to survive being dragged around the house and garden by a two-year-old!

The soft-hearted consumer may be put off by the consideration that in order to harvest an unbroken thread from the cocoon, the life of its occupant is terminated by steaming or baking. But my view, after some considerable thought, is that if we reject silk on this moral basis we should also reject eating lettuce; after all, snails and slugs will have lost their lives in the battle to coax this delicate comestible to maturity. (The silkworms are gathered and eaten, a valuable source of protein for human consumption.) Wild silks, on the other hand, are harvested from the cocoons once the industrious occupant has pupated, worked its way out of confinement and taken flight. Unlike their cosseted captive cousins in the silk houses, the wild moths seem to have more interesting prospects for life after pupating. Silk-producing moths are endemic across Asia, while globally there are also many cocoon-spinning species quite unrelated to the cultivated silk moth.

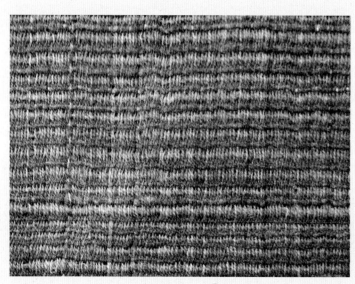

detail of a naturally dyed silk shawl from the Avani Kumaon project

In the Kumaon region at Uttarakhand in the Himalaya, the non-government organisation Avani supports craftspeople living in remote mountain communities by negotiating fair-trade prices for their work as well as assisting them in the development of alternative energy production to support village life and work. Wild silk for the project is harvested and sometimes combined with other fibres such as merino wool then spun and woven before being dyed with bio-regional natural dyes. The fibre is not as smooth and shiny as that of the cultivated variety; the diet of the wild moth is far more varied and the filament has of course been damaged in places as the moth broke through after its metamorphosis. The isolated communities supported by Avani can only be reached on foot and this textile-based project has increased their self-reliance while also creating work for the young people of the area, thus reducing their exodus to the already over-crowded cities of India.

The colour of wild silk fibre is determined by the diet of the silkworm and any over-dyeing is always tempered by the background shade.

Sugar cane

Sugar-cane waste (or bargasse) can be spun into a sturdy fibre. It's a good way to use a resource that is otherwise burned as fuel or made into paper, but the fine fibre dust can cause serious health problems for the production workers. In Okinawa, Japan, work shirts were being marketed as 'made from sugar cane' but the fine print revealed that only 8 per cent of the thread content in the cloth was actually from sugar, the balance being cotton. The Sugar Cane jean company in Britain also uses sugar-cane fibre in the making of their jeans, blending it with organically grown cotton, and applies a terroir-based classification to the various styles, which include Awa, Okinawa and Hawaii as the sources of the sugar cane or the indigo (or both).

Leather

Leather is produced from the skins of animals, sentient beings farmed by humans for food. Whether or not you endorse the wearing of leather is really a matter for personal judgement, but I am inclined to think that if the animals are already being slaughtered for their meat it would be hugely wasteful to simply compost the skins. Leather may be tanned using chromium salts, oils or vegetable tannins. The first are highly carcinogenic and best avoided; the third are likely to be rather more ecologically sustainable depending on how the tannin-bearing materials are harvested. Commercially, chrome tanning is far more popular as it consumes far less time than the vegetable-based process, which can take at least three months. But at least naturally tanned hides are less likely to poison the tanners or the region in which they work. The genus *Acacia* is a rich source of tannin, mostly from the bark. This also makes it a useful source of dyes for cotton.

Synthetic fabrics

The making of cloth from substances not necessarily evoking the thought of fabric – such as petroleum, wood, soy protein, rubber and milk – seems almost an exercise in alchemy. Nylon (a fibre initially developed in response to the silk shortages caused by the war effort in the mid-twentieth century) was quickly taken up by the hosiery industry at the conclusion of hostilities. It remained in use for all sorts of things, even bed sheets, until the 1970s when polyester fabrics began to fill its role. No matter how practical a non-iron, fast-drying fabric might be, when it makes the wearer itch and sweat or gives them smelly feet it's hard for it to remain popular!

this synthetic shade cloth has been destroyed by the elements it was installed to temper – sun, wind and torrential rain

Manufacturers of contemporary synthetics are seeking to improve the comfort and wearability of these fabrics by creating structures such as double knits that act to wick away moisture from the body and expose it to the outside air so that it can evaporate and not cause a problem by building up on the skin. Essentially considerable effort is spent on apeing the properties of natural fabrics such as wool and silk. One can't help thinking that investing that time and money in developing more sustainable means of growing and processing natural fibres would be far more beneficial in the long term. While polyester is promoted as a fully recyclable fabric, the energy used in recycling it is considerable; the harsh reality is that the percentage of fabrics collected and repurposed is relatively small and the end product is not another fabric but more likely simply a plastic object.

When such textiles end their useful lives in landfills they are slow to break down and indeed never decompose completely. One has only to look at tilled fields in India, for example, so see just how synthetics behave when they are discarded. The earth in India and other Asian countries is full of bright fragments of plastic bags, shredded synthetic twine and polyester fabric shreds as well as the ubiquitous cans, bottles, packing and discarded shoes.

Let me put my cards on the table. I have a deeply rooted aversion to synthetic fibres and avoid purchasing them. My sensitive nose can sniff out polyesters as well as picking the difference between wool, silk and cotton (especially easy when any of these materials is damp) and my extensive reading has instilled caution. I see no benefit in wearing substances that may cause cancer or that will make me feel uncomfortable when I can choose natural fibres that will breathe and that can be dyed easily with plants to make my clothes unique. Everything that is made in the world has an impact of some sort on the environment. We can choose to try to reduce our personal footprint by acquiring less, looking after it more carefully and thereby trying to make it last as long as possible.

even after many years, the brilliant red dye has run from this paeony embroidered on an *obi*

Dyes

At some stage after the yarn or thread is spun the fibre is also dyed or printed. Many of the synthetic dyes used to colour cloth are frankly toxic and when these are not completely rinsed away the remnants can induce forms of dermatitis. Some dyes may contain heavy metals, including antimony, cadmium, lead, arsenic, zirconium or mercury, to name but a few of the more popularly employed adjuncts. They might also contain formaldehyde, compounds of chlorine and bromine or other carcinogenic solvents.

the wear in this quite new cloth is likely to be due to the degradation of the fibre by synthetic dyes not being rinsed away properly

Traditional means of administering pattern to the surface of cloth include the intervention of resists – both mechanical (stitched, clamped or tied) and applied (wax, starch, mud or other pastes), which might be placed using stencils, drawn on with an applicator such as the tjanting or pressed onto the surface using a wooden block or metal stamp. The cloth might be dyed a number of times to produce various shades of colour or perhaps be discharged using some sort of bleach to remove earlier applied colour.

One of the versions of the stylish black so beloved of many fashionistas is achieved by the use of a dye commonly called 'sulphur black' that is assisted in the dye vat by sodium sulphide, sodium chloride (common salt) and sodium carbonate (washing soda) together with some form of lubricant. This hot dye process is followed by a wash with hydrogen peroxide and acetic acid and then a further process of 'polishing' involving yet more acetic acid. A crypto-anionic wetting agent might be used to remove any remaining loom grease and the process finished with a potassium dichromate bath.

Some of the so-called natural dyes of the past weren't much better. Consider this recipe for dyeing black with logwood:

The cotton is first left overnight in a decoction of 40 percent of sumach; it is then squeezed and treated for half an hour in a cold solution of iron pyrolignite followed by passing it through very dilute lime-water, and thoroughly washing. By this means iron tannate is produced on the fibre; the lime treatment serves to neutralize the acid. The iron pyrolignite may be replaced by 'iron nitrate' or ferrous sulphate, a little levigated chalk being added to the bath in the latter case. A handsomer black is obtained by the use of alumina mordants – a little aluminium pyrolignite, for example, being added to the iron bath.
The cotton thus prepared is next dyed with logwood, sometimes in the presence of a little fustic, by entering in the cold bath, which is then slowly raised to boiling. To increase the fastness of the black, the stuff may afterwards be treated in a bath of bichromate or iron nitrate. Sometimes the shade is darkened in the bath itself by an addition of copper sulphate.
Finally the cotton is soaped, a treatment that makes the tone of the black more agreeable.[16]

the tjap, a metal stamp for the application of wax to cloth

detail of a traditional wooden printing block, used in India to apply dyes and possibly also resists

dye samples at the Museum of Economic Botany, Adelaide Botanic Gardens and State Herbarium – although there is some doubt regarding the veracity of the label pertaining to the substance at the far right (the colour of the sample seems too lurid to be natural indigo)

There can be no doubt that both the modern method and the older one described on page 73 will generate quite toxic effluents. Dyeing with logwood (*Haematoxylon campechianum*) necessitates the felling of the tree to access the heartwood. This is not at all desirable no matter how much money it might contribute to a village economy.

Whatever means are used to dye our clothes, there is no getting away from the fact that somewhere energy is being used, resources depleted and effluents created. I prefer to take personal responsibility for these things and dye my clothes at home, using leaves from the garden, from windfalls or green waste collected from florists. In this way I am able to recycle the dye-baths as much as possible, avoid the use of toxic adjunct mordants and eventually return the waste materials to the soil without compromising my bio-region. Admittedly there is still a substantial footprint in terms of the heat energy used but in general I am reasonably confident that the methods applied will create less damage than those used by clothing manufacturers.

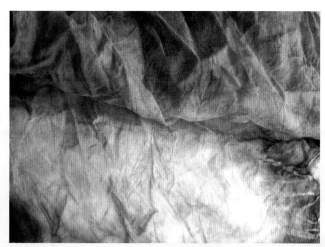

silk cold-dyed with plant extracts

◀ indigo field in the woods of Indiana planted by Japanese-trained American indigo master Rowland Ricketts III as part of an indigo research program

Other surface treatments

Some contemporary textiles have anti-microbial substances applied as part of the finishing process. Others have synthetic crease resistance added to them using chemicals that can later release formaldehyde vapours. The effluents from dyes are often contaminated with metals applied as metallic salt mordants to help fix or change colours or as a basic component of the dye itself. Such metals can include copper, antimony, tin, cobalt, iron, chromium, lead, zinc, mercury, cadmium and nickel. Fortunately studies are being conducted that reveal useful properties from a number of naturally available substances, such as plants and even milk. Endemic to sub-continental Asia, neem (*Azadirachta indica*) is a medicinal tree whose bark, stem, leaf, root and fruit can be used for anti-bacterial and anti-fungal action. The Milkymerino™ company in Australia is using a casein extracted from milk to stabilise and enhance the washing properties of fine merino wool.

Woolgathering

There are worse places to be than a woolshed. Especially on a cool day when it's just begun to mist light rain and there's the satisfaction of having all the sheep under cover. Shearers won't shear wet sheep and rightly so, as wet greasy wool gives off ammonia gas. Woolshed time is splendid thinking time. There are flurries of activity while penning the sheep and sweeping the floor and pressing the wool. There are also intervals during which one can stand and gaze at the grace and precision with which the shearer is divesting the sheep of its woolly coat.

English Leicester sheep

Milkymerino™ jersey dyed with eucalyptus and other windfalls ▶

chapter 3
Clothing choices

At the core of a fabric or garment's lasting usefulness is the idea of appropriateness.

Kate Fletcher[17]

detail of a *kantha* wrap, showing the wonderful rows of running stitch bonding the soft layers together. The word *kantha* simply means 'rag'

Essentially we want our clothes to be functional and practical as well as aesthetically pleasing, durable and simple to manage. But the concept of appropriateness goes a little further and embraces the relationship not only between garment and wearer but also that of the garment with its maker, ecological footprint and place. Considering these various aspects rather than making impulsive acquisitions will have positive implications for the planet as well as the pocket. Clothes must be appropriate not only for the occasions and activities for which they are intended but also in terms of their inherent worth, potential longevity and relationship with both wearer and maker.

the sari is an elegant, adaptable and practical article of clothing ▶

Japanese writer Sōetsu Yanagi, commenting on the use of pattern as ornament in the context of the hand-stitched indigo-dyed farmers' coats of his country, suggests that the mere use of pattern as ornament would be inappropriate, whereas the stout stitching used to join the two or more layers of the coats has intrinsic value in that:

> *Its charm is in its appropriateness to use and the strength of the stitching. The delightful patterning is incidental and utterly suitable. There is no concept of décor for its own sake. From this it should become clear that the origins of pattern are inextricably sewn into the fabric of use.*[18]

Certainly in the socio-economic demographic from which these garments originate each filament of thread and piece of fabric was precious. I imagine the women who were plying their needles would have taken great care to make their utilitarian stitchwork simultaneously beautiful as well as functional, thus bringing a degree of pleasure to what might otherwise have been a simply mind-numbing task.

During the Great Depression of the 1930s, many children in impoverished families wore clothes fashioned from flour bags. In the United States some flour millers caught on to the repurposing of their product and began to produce flour sacks with pretty floral designs to encourage frugal housekeepers to purchase their product over that of rival companies. Flour bags became clothing and were subsequently included in quilts.

The sari worn by the women of India is a wonderfully versatile garment. The length can vary from 5 to 8.3 metres (5½ to 9 yards), with the average being about 5.5 metres (6 yards). The width is simply that of the cloth as it comes off the loom – generally between 100 and 130 centimetres (40 and 50 inches). It is worn with an underskirt (into which one selvedge is artfully tucked) and a snug-fitting short-sleeved top known as a *choli*. One end of the sari is usually draped over the shoulder, but can also be worn to cover the head if required. It fits any body shape, is a modest garment that covers the body well and looks elegant. That it restrains movement and causes the wearer to take only small steps might be seen as a disadvantage.

As the garment is only shaped by the way it is arranged on the body, when it becomes too worn and threadbare it can very easily be utilised for other purposes, such as *kantha* quiltmaking, in which rows of decorative stitches are used to bond several layers of cloth together.

the sari fits perfectly on every body shape and size

When I was little, new clothes were usually bought or made because the others had been outgrown or had simply worn out. In my late teens I began to be seduced by fashion and fell into the trap of buying things simply because they were new and different. As an architecture student I began to construct garments that referenced the things I was learning about building design. I thought about shape and form and about ways of cutting the cloth for least waste. I made boxy jackets that bore relation to kimono and featured stacked pockets reminiscent of Habitat, Israeli architect Moshe Safdie's apartment building for Expo '67 in Montreal, Canada. I designed skirts that were based on the sarong, that efficient rectangle of cloth that transforms to mould to the body of the wearer; wide tubes that could be folded to hug the hips and then simply tucked over twice at the top to stop them falling down. In the early 1980s I took up a position as a trainee exhibitions curator at the Adelaide Festival Centre Gallery, was exposed to 'wearable art' and promptly fell under its spell, making a series of wildly coloured (and highly impractical) garments. I knitted some frankly ridiculous jumpers, one memorably based on the patterns formed by seaweed when cast up on the shore. Being slightly dyslexic I found that knitting patterns would simply morph into a blur of jumbled hieroglyphs before my very eyes and so I blithely knitted forth without formal guidance, casting on and off and sometimes knitting whole handfuls of stitches together to create a garment that more or less covered the vital bits. It was about this time that my father began to be embarrassed to be seen in my company. Half a lifetime ago I would even wear lycra; the thought now makes me shudder with shame. Lycra is, after all, a fabric that clings to the body, revealing far too much (a little mystery goes a long way). Things are very different now. These days I eschew synthetics completely, wear at least 50 per cent repurposed garments and do a lot of mending. My favourite fabrics are wool, silk, linen and hemp with the occasional inclusion of organically grown cotton.

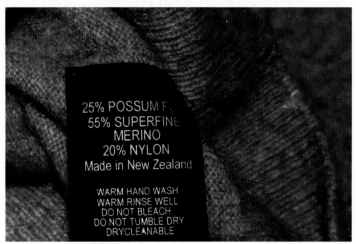

the swing-tag on this garment read 'all natural' but the inner care label reveals the truth

For me one of the most important factors when choosing clothing is the primary substance used in the garment. Remember that it will be in contact with your skin and tiny particles that are inevitably shed from the cloth (whether fibres, gaseous fumes or potentially carcinogenic dye dust) are likely to be absorbed by it or at least affect it in some way. Hold the cloth up to your nose. If a scent can be detected then the cloth is actually 'off-gassing' and those vapours can potentially compromise your health.

The other significant consideration is the provenance of the garment, not only in terms of the materials but also in regard to where it was made, how and by whom.

The tag exchange

Lea Redmond of Leafcutter Designs has designed this tag as a way to introduce and cultivate the kind of imagination and awareness that is a prerequisite to ecologically and socially conscious shopping habits. In exchange for one (or more) clothes tags, Lea will send one (or more) of these tags to be sewn into your clothes as a reminder of the need to be aware of where your clothes come from and how they are best treated. As the garment is passed on to family, friends or thrift stores, the tag carries and spreads the message. www.leafcutterdesigns.com

a tag to think about

The quality of the garment is another most important factor. The fashion and textiles industry consumes vast resources, among them many that are finite. Huge volumes of water as well as shocking quantities of chemicals go into producing cloth. It makes good sense to choose a garment that is well made and well fitting, comfortable, resilient and appropriate for the kind of wear that you are expecting from it.

organic cotton jeans dyed in Japanese indigo: Denham the Jeanmaker

wonderful organic merino socks-in-a-box hand knitted by Jo Kinross of Nelson (New Zealand) have toes reinforced with silk for longer life and come with matching yarn for repairs

Making informed choices

What are the criteria that you consider when shopping for clothes – is it the quality of the fabric? Or perhaps it's the cut of the garments? Does the sturdiness of the stitching attract you? Do you think about how they might survive the ordeal of the laundry? Is it whether they epitomise the fashion of the time or whether they are timeless in style?

The sad truth is that in the age of fast fashion many of us have fallen into the habit of buying things simply because they are cheap and available rather than out of necessity or with a clear buying plan. In those throwaway decades of the 1980s and 1990s, people were blatantly encouraged to stock up on a pile of new T-shirts so as to simply peel another one out of the packet when the one they were wearing became grubby. The truly cool simply threw their dirty clothes in the bin.

Now in the second decade of the 'second millennium' we are beginning to acknowledge that our spendthrift ways cannot last forever and that just as in the Great Depression and during World War II, when things were carefully cherished and made to last, it is now time to again pay more attention to our clothes, not only in how we treat them but in how we choose them as well. While we might not face critical fibre shortages at present, the energy needed to grow, harvest and process fabric and to construct clothes is becoming daily more expensive (both economically and in terms of the cost to the environment).

dressmaker and garment repairer Amy Oakes (New Zealand) sewing at Aeon, her atelier in Cuba Street, Wellington

Many naturally sourced fibres receive extensive applications of potentially toxic chemicals during the growing period as well as during processing and this cannot be a good thing. The synthetically produced textiles that were welcomed in the twentieth century for their ease of management and wrinkle-free properties have their own set of challenges. Although many synthetics are admittedly recyclable, the manner of reconstitution is an energy-expensive one; firstly in terms of collection and secondly for the process itself. In this age of rapidly diminishing oil reserves it might be sensible to use that substance for fuelling vital transport rather than churning out more plastics –and when it comes to the crunch, polyesters and plastics never really decompose properly. On my visits to the sub-continent I have observed with some horror the shreds and fragments of plastic that are now to be found scattered throughout the country. Every tilled field is full of coloured remnants from clothing and packing materials. It would appear that not a single handful of earth is free of this scourge except perhaps in the deepest jungles.

Some might argue that recycled fabrics would be a responsible choice. This may sometimes be true, such as in the case of Japanese *saki-ori* (rag weaving). But what about the polar fleece purportedly created from PET bottles? The popularity of this insulating substance has encouraged some makers to bypass the collecting of pre-used plastic drink bottles and to simply take delivery of newly made bottles directly from their manufacturers for metamorphosis into cloth. This makes no sense to me at all.

Just because material is recycled does not make it automatically benign, especially if it was not designed specifically for recycling.[19]

The current predilection for adding the prefix 'eco' to labels as a way of appealing to a potential market[20] often amounts to blatant 'green-washing'. It's as if giving it a 'friendly' label will make everything right. Certainly it is a clever marketing ploy. A T-shirt might be made from organically grown cotton, but how can it possibly be ecologically sustainable if it is brightly coloured with synthetic dyes? No synthetically sourced dye is truly safe and those who state that these colours are harmless are frankly lying.

The matter of the health and wellbeing of clothing workers at every stage in production is also a major cause for concern. We are so far removed from the origins of our clothes that we have no idea how the growing and harvesting of the fibre or the processing of the cloth affects the workers. But the stories that filter through the media together with some of the exploitation I have seen at first hand offer sufficient dots to join so that a picture can be derived. Workers in developing countries receive far lower wages and work longer hours than their Western counterparts. The conditions in the factories are often appalling and the emissions to the surrounding environment substantial. Even the revered William Morris, who passionately promoted the use of 'natural' dyes more than 150 years ago, was responsible for polluting the waterways of Britain with the waste from his processes. Water that entered his dye house clean was flushed away laden with mordants and unbonded dye residues.

In 2007 the Taihu Lake in Jiangsu Province, China, was found to be so disgustingly filthy that even the government could no longer ignore it. The water was fluorescent green with toxic cyanobacteria, there were dead fish accumulating on the surface and the whole lake reeked abominably. Unfortunately, as well as being the de facto sewer for the factories of the region, it was also the source of the local reticulated water supply and of shrimp and fish. Thousands of rice farmers relied on the lake for irrigation. The problem had been gradually building since the region began to be industrialised in the 1950s. The dye houses that were established there enjoyed the convenience of using huge volumes of water pumped from the lake and then simply flushing the waste back into the same system. The lake's eventual death wasn't a matter of 'if' but 'when', and lest the gentle reader is lulled by the thought that this disaster happened in another country, think again. Anyone who has purchased a cheap Chinese clothing item (and who among us is innocent?) has been a complicit contributor by association.

The textile industry in Australia, where at least some laws are in place to offer guidance about work practices and emissions, has been largely destroyed by companies removing their factories to countries where there are fewer or no controls and fewer expenses. DPK Fabrics in Sydney set up a state-of-the-art dye and knitting house where the effluents were recycled, the dyes applied in closed systems and the heat energy from processes harvested

◀ Emma Christie (Australia) wearing a ruched and tattered silk skirt that simply becomes more beautiful with age

and reused, investing millions of dollars in plant development. Only months later a series of hiccups in the world economy saw this company close its doors and take its business to Asia. When things like this happen it should be a huge concern, not merely for the environment but also for the loss of employment and skills in the host country. Eventually the world will run out of oil due to excessive consumption and irresponsible harvesting practices. Not only will we have to do without luxury foods shipped across the oceans, we may see ourselves once again personally responsible for the clothes that go on our backs and the shoes for our feet. Those who joke about knitting sandals from hemp to save the environment might find themselves doing just that, as the once thriving textile, clothing and footwear industries of the nation have long since been consigned to history.

a welcome sign of the times, local dressmakers are once again increasing in number

Clothes made in Australia certainly cost more even without trans-oceanic shipping, but an investment in locally made products is also on many levels an investment in the future, and so amounts to a socially responsible act.

A little reflection on the relative price of the garment on the rack should give us an idea of how the funds are distributed, even if we don't know much about the conditions under which the makers are working. Let's start with the retail price of a dress, perhaps made of a mixture of silk and cotton with a little decorative embroidery, and try to break down the costs. The floor price at the store is always double the wholesale price. The fashion agent and the designer have the next biggest slice of the pie. Subsequent expenses include the materials and trims from which the garment is made followed by the packaging and labelling of the item. The smallest morsels account for the makers sewing the garment and the workers producing the resources from which the actual cloth is made. Food for thought indeed.

When we look at simple garments such as T-shirts the figures are sometimes quite horrifying. When the wholesale price of a cotton top is less than one Australian dollar at the point of shipping, including packaging, tags and a 15 per cent profit margin, it's clear that huge volumes have to be processed for money to be made. Invariably management will be collecting the cream of the income so the position of the workers on the factory floor (or worse still, in their own houses) cannot be enviable. Cotton growers in Asia, already battling to make money from their crop, are finding that insects are becoming increasingly resistant to pesticides, which in turn are increasing in cost. Consequently suicide rates are on the rise as increasing numbers of small-holding cotton farmers despair, fearing the consequences of the failure to keep up loan repayments when harvests fail. Growing organic cotton takes the same amount of water and land but requires more intensive labour in terms of weed control and pest management, so we should be prepared to compensate the cotton farmers appropriately.

General wastage in the clothing industry is also rife, not only in the form of loom-ends and cutting surplus. There have been instances of fashion chains dumping entire clothing stories in the garbage, first deliberately damaging the clothes so they are less likely to be worn. Haute couture fashion designers have been known to burn unsold collections rather than see them sold off cheaply at the end of the season.

Custom options

A number of designers and clothing merchants use the internet to their advantage and with accompanying environmental benefits. Customers choose the items they want from a range presented in a virtual store on the designer's website and then place an order, after which the desired garment is sewn specially for them in the colour and size specified by the client. Admittedly this removes the purchaser from the social and tactile experience of shopping, whereby garments can be felt and tried for fit, but it means that unnecessary items are not produced and reduces the need for storage space. It can also save fuel and money on travel, but clearly this option will not appeal to everyone. Delivery certainly takes a little longer than the instant gratification of buying a cheap T-shirt at some fast fashion chain, but there's also something quite charming about the anticipation and the waiting for the parcel to arrive.

By making each piece in response to a firm order, the company saves on costs associated with storing finished stock and does not have to make guesses at the number of garments, sizes or colour ranges to have on hand. The process takes more time but ultimately saves on energy and materials.

A well-known sneaker company offers a service whereby potential clients can play on their website and customise a pair of sneakers in their own size. They present a wide range of colour choices as well as a selection of prints and will even personalise your sneakers by adding your name (or whatever other tag you'd like). While the materials and colours used might not be ecologically sustainable it is at least likely the customer will value and care for their shoes more because they will feel that they have participated in the design process. On the other hand, as far as I can determine, the company employs cheap under-age labour, so they still have some way to go before being worthy of endorsement by those with a social conscience.

The next significant issue to do with clothing is the manner of maintenance. An item of clothing uses more energy and water in a lifetime of wearing than is consumed in its making and this is where the makers of synthetics may have a marketing point, as their product may well require less cleaning and pressing. On the other hand, the myths and legends surrounding synthetic fabrics are many. It is a popularly held belief that polyester fabrics are more environmentally friendly than natural-fibre fabrics because they can be recycled. But the catch is that they can only be recycled once[21] and then not into cloth, as is so often suggested, but into such things as garden stakes, pots for plants and parking bollards. Inevitably, though, these items will also break up into smaller fragments, but they won't biodegrade or become reabsorbed into the ecology in a useful way. This means that no matter how you look at it, synthetic materials will have a negative impact on the environment for a long time.

dyed fair-trade organic cotton T-shirt individually dyed using plant material and rusty nails ▶

So in a nutshell, the things we need to think about when acquiring clothing are:

What is the garment made from?

Who made it and where?

Were they adequately recompensed for their work?

How far has it been transported?

How has it been coloured?

How well is the garment made?

How long is it likely to last?

What will I need to do (and how often) to keep it clean?

Could I make it better and more beautifully myself?

Is it appropriate (in every sense of the word)?

Do I truly need it?

It might also be worthwhile considering whether the garment really needs to be brand new or whether it might be cheaper (and safer) to buy it second hand. Opportunity shops and thrift stores can save us a great deal of money, put discarded clothes to good use by giving them an extra lease of life as well as contributing to your own distinctive and fabulous style and generally supporting a charity as a bonus.

The value of 'less'

The less we acquire and the more carefully we manage it, the less has to be produced to satisfy demand and the less we have to dispose of later. If you want to have a small footprint on the planet, have less 'stuff'.

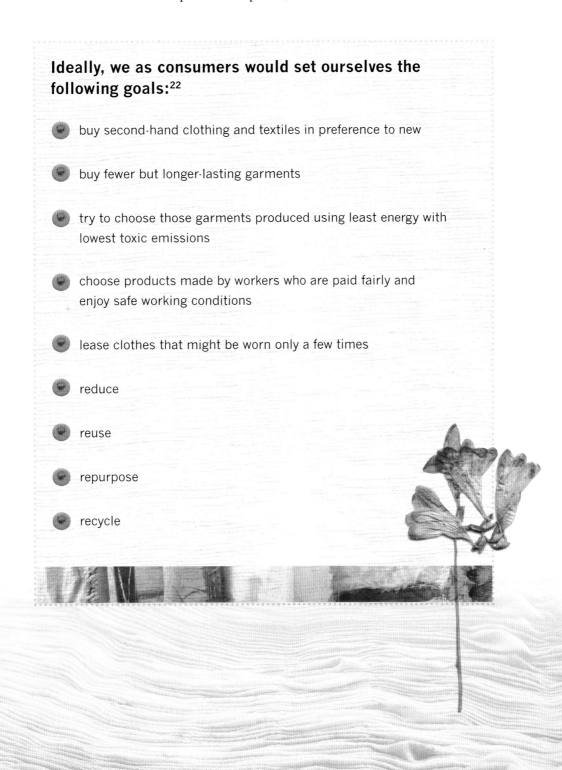

Ideally, we as consumers would set ourselves the following goals:[22]

- buy second-hand clothing and textiles in preference to new

- buy fewer but longer-lasting garments

- try to choose those garments produced using least energy with lowest toxic emissions

- choose products made by workers who are paid fairly and enjoy safe working conditions

- lease clothes that might be worn only a few times

- reduce

- reuse

- repurpose

- recycle

chapter 4
Planning your wardrobe

It is up to date to be resolutely no longer up to date

Barbara Vinken [23]

Travelling light

Years of travelling have taught me just how few items of clothing are really necessary in order to get by. As a child of ten, back in 1968, I had a relatively limited wardrobe especially when viewed in the light of today's expectations. There were two shift dresses (hand sewn for me as a Christmas present by my doting aunt), a pair of slacks (I wasn't allowed jeans in those days), a couple of skivvies and a jumper hand knitted by my mother. I also had a couple of 'Sunday best' dresses so exquisitely sewn (again by my mother) that I had to be extra careful when wearing them and consequently (given I was a hopelessly clumsy child prone to climbing trees and falling into puddles) often spent special occasions quaking in the fear of wreaking accidental damage.

During the week I wore a school uniform. In summer I alternated two gingham dresses and in winter sported two blouses and a ghastly box-pleated tunic that was to all intents and purposes made of the same evil coarse wool as the habits worn by particularly penitent nuns. (Some years later at another school, the box-pleated horror was replaced by a kilt, rather more fun to wear and equipped with a handy pin for guerrilla attacks on fellow students.) When we left Australia for a year of world wandering – my father being on sabbatical leave – I simply stuffed the lot into a small suitcase, sans uniform of course, but with the addition of the necessary socks and scanties. As we were travelling into the European winter I was also given a red wool coat (because as everybody knows, red coats are warmer). Woolly stockings, a warm hat and mittens completed the picture. Over the next nine months I happily wore these things in various combinations and often in multiple layers. When at length our journey took us to America, I acquired a stunning pair of turquoise Bermuda shorts together with a peaked cap made from emerald green velvet, both discards from a relative. I was much taken with them and wore them continuously, together with an aquamarine skivvy, until the first snows of winter made the wearing of shorts somewhat impractical. But I still insisted on wearing the cap. In combination with the red coat it must have been a startling sight.

my mother and grandmother in clothes sewn on the old Singer

I went through various clothing obsessions as a child. Not satisfied with being the only honey-coloured child in school at a time when neither indigenous nor Asian students graced the portals of the institution I made myself further conspicuous by my clothing choices. After seeing the film *Lawrence of Arabia* in my early teens, I sewed a voluminous white garment from an old sheet and proceeded to wear it everywhere, even to school on 'casual' days. I'm surprised I wasn't arrested. Some years later I managed to negotiate a purple velvet kaftan from a fellow student who was selling them at school on behalf of his uncle and horrified my parents by wearing it to the theatre, barefoot. For a while this garment became my uniform unless I was on horseback, when I would revert to jeans and a T-shirt (but still barefoot).

my mother and me

On a trip to Europe when I was seventeen, my mother decided I would need a rather more formal wardrobe. With her vision substantially impaired by the particular rose-tinted ocular enhancement pertaining to motherhood she fondly considered I had a passing resemblance to the eldest child of the then Princess of Monaco and had an idea I should be garbed in smart suits rather than my usual hippy tat. She made me a fabulous grey tweed wool trenchcoat with lots of buckles, flaps and pockets. It had a matching skirt that kept the winter chills out nicely and I wore them frequently. But the peacock blue silk satin suit, sprigged with printed gold flowers (gentle reader, I shudder as I write) that was packed into the suitcase with its hem still pinned (but not stitched) never even saw the light of day. Its one moment of glory was the moment it accidentally stabbed and drew blood from the fingers of a Malaysian customs officer. This gentleman had (after delving happily in our luggage) held up a pair of my unmentionables for general amusement but leapt back in horror after a deeper investigation – doubtless in search of more entertainment – resulted in a smart jab from the unfinished masterpiece. I expect he may have thought he'd been bitten by something. In any event his approach to the next piece of luggage was a little more circumspect.

My mother's dreams of my sartorial splendour were further dashed as I spent the entire summer romping about in a rose-print skirt and peasant blouse, picked up at some market or other. Looking back there was an obsessive–compulsive aspect to my dressing that emerged fairly early on and has stayed with me ever since. Colour fixations have come and gone.

I began with a serious commitment to orange and blue (not exactly subtle, but hey, I was only five) later followed by a passion for purple that was subsequently eclipsed by more sombre hues. There was a decade during which I wore nothing but black. At the time I worked in arts administration, and artists and administrators ubiquitously wore the colour like a bevy of professional mourners. Rumour has it that the wearing of black was brought into fashion by New York's theatre mechanists. They wore nigrescent shades so as not to be spotted while shifting scenery and moving props during scene changes and simply left the clothes on when they wandered out to play at the end of a shift. Other New Yorkers picked up on the habit, gradually the style caught on and it remains popular, particularly among arts and fashion workers. The fashion editors who dictate the 'in' colours and styles months in advance – and are in all probability completely over them by the time their impressionable readership sits down with the latest edition and a cup of tea – also seem to sport standard black as their uniform.

These days I am somewhat limited in my colour choices. I have learned too much about how clothing is made, the chemistries involved and the consequences of garment production to feel comfortable buying synthetically dyed clothes. I now try not to buy black as the dyes used are among the most toxic of the synthetics. On the rare occasion I do purchase such things they are either pre-used or of exceptional quality so that I can be sure that they will give plenty of wear. This very much narrows the field when choosing garments and fabrics.

After some fifty years on the planet I've managed to trim the quantity of clothes I travel with to a bare minimum. This leaves more space for essentials such as a small cauldron for on-the-run dyeing of windfall bundles, notebooks for writing and drawing as well as the indispensable sewing kit. It also leaves space for the inevitable pocket stones and the as-yet-unfound treasures that I like to bring home. A dogspike salvaged from the railway near Picton, a small metal utility cover from the streets of San Francisco, the London copper scrap and irresistible old garments from op-shops and flea markets will all need luggage space eventually.

My travelling wardrobe now usually consists of a hemp shift dress with big pockets, a slightly shorter wool shift, a pair of drawstring linen trousers, a long-sleeved T-shirt and another with three-quarter sleeves, a couple of stretch-knit crop tops (in lieu of more structured brassieres), four pairs of knickers, two pairs of socks, a loose wool cardigan, a big silk or wool scarf, a soft beanie, a pair of soft silk drawstring trousers and a long-sleeved silk top. The silk pants are worn for sleeping but can be layered under the linen pants in cold weather. Similarly, the wool shift can be worn over the hemp shift. If I'm travelling to a really frigid climate I also take my coat (made from two repurposed wool blankets), a pair of organic cotton tights and a pair of wool legwarmers. I always take an apron, sometimes two. All of these items have been dyed with plant dyes and their colours seem to relate quite well; also the variable leaf prints on the fabrics mean that any accidental stains don't usually look too bad. Nothing ever needs ironing and if required I can wear everything at once to keep warm. I alternate the T-shirts and wash them along with the daily unmentionables every evening. It's a low-maintenance and practical wardrobe that doesn't take up too much space.

How many clothes do we really need?

The answer is very much a matter for the individual. Actively planning your wardrobe in relation to your profession and the sorts of activities that feature in your life is only common sense. Who hasn't fallen into the trap of impulse buying a garment only to find that it doesn't go with anything else, falls apart at the first wash or simply screams 'hag' when we stand in front of the mirror? I never shop when depressed or sad – it always ends in disaster. Either the mood simply deepens with the horror of the fitting room or something hideous is likely to make its way home under the guise of being an object of solace, only to be revealed a short time later as a changeling with nasty habits. Opportunity shops, wardrobe depths and thrift stores are full of such items cast off by people who have awoken to the refrain 'What was I thinking?'. This of course makes for rich pickings if you're looking for materials to repurpose, so it's an ill wind that doesn't blow somebody any good.

Even though I find the observation of changing fashions quite fascinating, the older I grow the less I enjoy shopping as a pursuit. While every now and then I wander foreign streets investigating emporia (sometimes, I blush to admit, in lieu of museums), I tend to scan the rack in search of interesting detail rather than with intent to make a purchase; so I almost fell into hysterics while scanning an exhibition catalogue that described the wardrobe philosophy of the person who had donated the particular clothing collection to the museum. The essay explained her very disciplined habit of limiting her apparel selection to only twenty garments, retiring an item to storage when a new one was acquired. I would be hard pressed to find ten dresses in my closet, much less twenty.

Internationally renowned designer Coco Chanel not only revolutionised fashion, she is popularly thought to have epitomised frugality in her clothing. Legend has it that at the time of her death she had only three suits hanging in the wardrobe of the suite she occupied at the Ritz in Paris,[24] but perhaps that was merely her basic travel kit. Certainly the clean and practical lines she developed made life much more comfortable for Western women. Chanel

Romance was Born (Australia) from the exhibition 'Fashioning Now', 2009: this label is known for its eclectic 'bowerbird meets tailor and bricoleur' approach to fashion

is also reputed to have shrugged off the possibility that her copyists might prove an economic threat and apparently much enjoyed watching her style spread like wildfire through all levels of society.

It is obvious that the ease with which information is disseminated around the world has had rather a large impact on the way we dress and behave. The existence of the internet means that we can almost instantly be aware of trends and fashions overseas, a far cry from the days (a mere twenty years ago) when the latest fashion magazines were eagerly awaited as the source of such information.

installation view of the author's plant-dyed garments in the 'Planeta' exhibition, Møstingshus, Copenhagen, 2010

Trend forecasting is big business. People such as Lidewij Edelkoort and her company Trend Union specialise in analysing popular culture globally and then making predictions as to subjects, activities and colours they think will be in demand. Their 'look books' are beautifully devised and include series of themes, silhouettes and stories as well as details of heel styles, skirt lengths, pant shapes and accessories. But trend forecasting doesn't stop there; it wanders on to embrace almost every facet of living, from architecture through to the desirable shapes of cars. It doesn't take much to recognise that these predictions will be largely self-fulfilling, as most manufacturers and designers would be taking note even if they might not publicly admit to it. How else could purple suddenly be 'the new black' on both sides of the Atlantic, or the kaftan make its surprise return? Or perhaps there is some other form of collusion that I as a mere mortal have not yet discerned. Either way it is always entertaining to muse on what might happen next. The fashion for short hair invariably morphs into a craze for long hair, while square- or round-toed shoes follow winklepicker outlines.

A degree of comfort can be gained from running with the pack. As the saying goes, when in Rome, do as the Romans do; meaning that adapting oneself to local standards can make life a little smoother. We take off our shoes when entering temples and cover our heads when entering Roman Catholic churches. We wear black when attending funerals and avoid wearing white at weddings. Wearing a swimming costume at the seaside is accepted, but that same item of clothing would be considered inappropriate to wear in the street or when visiting the supermarket. A former Premier of South Australia who wore pink shorts to parliament in the 1970s was merely made fun of; in some political situations, conforming to the norm can mean the difference between survival and possible demise.

China's Chairman Mao excised fashion entirely from his country by imposing on his compatriots an austere uniform comprising a simple buttoned shirt together with a pair of trousers. While this may have been partly as a means of (at least initially) ensuring the fairer distribution of limited textile resources so that each person had a basic outfit, it also had the politically useful effect of subjugating personal expression in dress and of melding individual members of

◀ intricate lacework, timeless and beautiful

Latvian traditional clothing. That's my grandmother at left in the front, photographed in the early 1940s. At the time, Latvia was under German occupation, so these women (all bridesmaids at a wedding) had made a point of wearing traditional garb as a subtle sign of resistance

the population into a more easily managed collective. Ironically, 'Mao shirts' enjoyed popularity for a time in the capitalist world, worn as a fashion item.

In 1746 the Hanoverian government of Britain passed an act of Parliament restraining the use of Highland dress, which banned the wearing of the kilt as a means of controlling the Scots, although curiously the ban did not apply to the military or the upper classes. The ban was lifted in 1782, kilts were donned again and tartan began once more to rise in popularity. Eighty years later Queen Victoria and her consort Prince Albert each designed a tartan for their home at Balmoral.

The little brown dress project

Wearing a uniform or some specifically designated garment during working hours can of course reduce wear on the rest of the wardrobe. Dull as the thought of being regimented might be, it can certainly save having to make daily decisions about clothes.

Some years ago an artist in Seattle, USA, decided that she would spend three hundred and sixty-five consecutive days wearing one 'little brown dress'. For a year she wore the dress every day as a very public statement against the sort of consumerism that requires us to constantly buy new clothes. In winter she added other layers, including knitwear and trousers, while in summer she wore it bare legged; essentially the accessories changed but the dress remained a constant. In our fast-fashion society, where it appears that every day must have a new look, this initially seemed quite a bold venture. Amusingly, though, the artist began to observe that other people regularly in her orbit did not appear to notice that she repeatedly wore the same item. Her conclusion from this was that most people are too preoccupied with themselves to worry too much about what others are wearing. In general I am inclined to agree, although obviously if you turn up to a black-tie event in a surgical gown somebody is bound to notice sooner rather than later. The existence of the internet allowed the artist to publish her images of her combinations and no doubt provided a certain amount of encouragement as viewers from around the world offered comments, suggestions and support.

Of linen, lace and leather

Irish immigrant Daisy Bates – who became infamous not just as a bigamist but even as a 'trigamist' and then achieved further notoriety by doing exactly as she pleased – eventually took up residence on the East–West Railway line that traversed South Australia heading for Kalgoorlie and Perth. She set up camp in a tent at Ooldea. Here she intended to 'smooth the pillow' of the indigenous population as they passed, believing (wrongly as it happened) that they were destined for extinction, while also attempting to prevent them from being corrupted by the influence of those connected with the railway line. Despite the extreme temperatures that prevailed in this region, Mrs Bates insisted on keeping up appearances and dressed in the formal

Edwardian style of the day. Some of her clothes are kept in the vaults of the South Australian Museum, where I was privileged to view them some years ago. In preparation for desert life she had her skirts made with a small strip of leather affixed to the inside of the hem so as to prevent fraying. The underarms of her close-cut woollen jackets were fitted with small pads to absorb perspiration and prevent stains from being visible on the exterior of the garment. The exquisitely tailored suits had been carefully and extensively mended with tiny stitches, as had her blouses. There was also a small collection of paper patterns, cut from newspapers, letters and pieces of cloth. Daisy Bates let nothing go to waste.

It's interesting to compare the wardrobe sizes and storage space considered desirable in these times with those of yesteryear. Older closets are much smaller and also narrower across the part that accommodates the shoulder of the garment. Clearly as a population we have grown bigger as well as having a lot more clothes. Visiting display homes, we find that even working-class families apparently now need walk-in wardrobes and where these are not provided the storage space simply takes up entire walls.

One of my favourite books as a child was *Little House in the Big Woods* by Laura Ingalls Wilder, a largely autobiographical story about growing up in the woods of Wisconsin in the 1870s. It was a rich but hard-working life in a one-room log cabin, built around the things that had to be done as a matter of self-sufficient survival in those times: planting, gathering, harvesting and preserving along with mending, making and the quiet religious observance that made some Sundays seem very long indeed. When I first read the book I thought that spending the seventh day of each week being quiet sounded purgatorial; forty-five years later it seems heavenly.

It was a way of life that puts the excessive consumption that marks our times into rather harsh perspective. The old Jewish proverb 'Who is rich? They who have enough' encapsulates the spirit of life in the little house neatly; whereas we are constantly being adjured to buy more, eat more, build more and are encouraged to believe that material possessions will contribute to our worth as people. The reality is that these things simply burden us with debt, objects and responsibility.

woollen baby jumper mended, embellished and eco-printed with *Eucalyptus cinerea* leaves

A return to a slightly 'smaller' life,

being responsible for fewer garments but ensuring

what we do have is made with skill and kept with care,

would leave us much more time to

enjoy simply living our lives.

chapter 5
Making clothes

Can you make me a cambric shirt?

parsley, sage, rosemary and thyme

without any seams or needlework

and you shall be a true love of mine

traditional English song

I don't think I'll ever forget the sense of wonder and delight I felt when my mother first showed me how to join two pieces of fabric together using needle and thread. The process of placing two pieces of cloth right sides together, sewing them along the edge and then turning the object right sides out to show a smooth transition between one piece and the next seemed simply magical. I remember turning the work over and over, for even though I had watched her sew clothes countless times it wasn't until I had done the stitching myself that it really sank in. At first I used simple running stitch, working it in one direction and then back again. Then I progressed to back stitch, although the word 'progress' is perhaps too strong, as my thread seemed to get fearfully tangled. The simple lines of running stitch are still my very favourite.

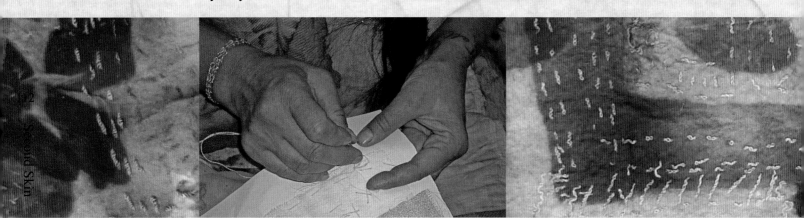

As a child I was keen to experiment with the sewing machine and I must have driven my mother to absolute distraction until I finally understood that if I didn't hold the threads aside as the machine was set in motion they would be devoured by the cogs underneath. Inevitably they tangled again, I would demand assistance and mother would grumble in from the garden where she'd been happily delving in the earth among the roses, rinse her hands and apply herself to solving the problem.

On the days I was at my grandmother's house I was sometimes permitted to use her precious hand-cranked Singer sewing machine. While I perched at the table squinting at the cloth and attempting to coordinate my wheel-turning hand with my fabric-guiding hand, my small brother would sit astride the wooden sewing-machine cover on the floor pretending to be a cowboy. It's a miracle the machine has survived so long.

When my grandmother was faced with leaving her Latvian homeland in 1944 she had a scant hour between learning of the last train out and its anticipated departure time. To this day I am still in awe of her composure as she made swift decisions about what to take and what to leave behind. In the short space of time available she organised her three children and their clothing, buried the best silver in the garden (hoping to be back sometime), packed the second-best silver so as to have tradeable items in lieu of money and found food for the journey. (She also packed her Sunday-best dress that had been made for her in the 1930s, and some fifty years later wore that dress to my wedding.) Amid the turmoil her calm good sense caused her to take the sewing machine. This miracle of precision engineering had been purchased by her in 1927 and had already earned its keep well. It continued

to do so in the years that followed. How my tiny grandmother managed to schlep the solidly constructed sewing machine while keeping tabs on her children and dodging the bombs is a mystery.

I eventually inherited this gorgeous treasure and it still sews beautifully today. Being hand powered it can be used anywhere there is a surface to put it on, the only slight disadvantage being the potential for developing huge muscles in one's right arm from a surfeit of enthusiastic cranking. My grandmother knew her body so well she could lay a piece of cloth on the floor, take a pair of shears and cut it to size, sew it together and confidently expect to wear it without the need for fittings or alterations. Her colour choices were often somewhat eclectic and sometimes included luridly patterned offerings from the remnant boxes at Melbourne's splendid Victoria Markets. The hippy era was in full swing at the time and made memorable by some extremely loud floral fabrics that unsurprisingly caught her eye and (more remarkably) found their way into her sewing basket. Grandmother would cheerfully mix and match, declaring that if the colours were 'good enough for God in the garden' they would do just as well for her.

Her clothes conformed to a very simple prototype; a sleeveless shift for the very hot days of summer and the same style of shift but with the addition of three-quarter sleeves for other weather. Over the top there was usually an apron of some kind, enriched with a number of pockets and tied up behind with a simple bow. Underneath she wore simply splendid silk bloomers hand stitched with French seams and elegant lace trims, suspenders and stockings. When she needed extra layers for warmth she would drape a selection of cardigans and shawls over her shoulders (the latter always with a small concealed pocket for a hanky or other necessaries). The cardigans were as colourful as her dresses, knitted or crocheted from either remnant or recycled wool yarns. She grew her hair long and wore it braided and pinned up in a lovely soft spiral bun. As a rebellious teenager I wondered that she always kept to the same mode; now I too conform to a similar style, the difference being that I wear trousers under my dresses rather than stockings.

My mother, on the other hand, favoured far more fashionable clothes. She made do with minuscule housekeeping money, was the absolute queen of 'a hundred and one ways with mince' and grew vegetables in the garden as well as spending her summers working from dawn to dusk preserving fruit from the fifty-two highly productive trees in our back garden so as to be able to put money aside for beautiful cloth. This was mostly purchased from the remnant box as well, so she had to do some rather clever cutting to stretch the material as far as possible. She developed a number of variations on a Coco Chanel-inspired suit pattern and my favourite was a slim gold shot-silk dress that through the miracle of the weave was brilliant scarlet on the inside. It had exquisite piping around the waist created from sewing some of the silk around an old pyjama cord and was complemented by a matching short jacket. She even fabricated a stylish clutch bag from the same material, using cardboard from an old cereal box for stiffness and cunning to cobble it together.

Trial and error

When attempting to sew my first shirt according to a pattern from an old red velvet curtain that I purloined from the store cupboard, the process of turning a sleeve absolutely baffled me. Eventually I swallowed my teenage pride and went and asked my mother for help. I must admit to still finding it sometimes difficult to comprehend how a sleeve with the right side out can be inserted into a garment with the wrong side out. I understand that the process works but have trouble explaining why. Dyslexia doesn't help at such times!

I attended primary school in the 1960s, when formal sewing classes were still part of the curriculum. We stitched away at samplers, sewed peg aprons and made ghastly Humpty Dumpty cushions filled with beans. These were much appreciated by the rats that crept into the house during the winter months. We also sewed some rather bizarre dresses from calico that we first undertook to colour using an all-purpose dye powder. Looking back, our class teacher must have been a hippy at heart as she guided us in the basics of what was then referred to as 'tie dye' but is more properly known as *shibori-zome*. The process of first hemming the calico, then hand washing it, ironing it and then adding resists in the form of rubber bands took several weeks of class time. By the time we had completed the sewing on the vilely patterned cloth after laboriously adding a myriad tailor's tacks, eyelets and an impeccably stitched hem, an entire term had become but a memory. Many of us had enjoyed a recent period of physical growth and could only fit with difficulty into our laboriously produced garments. At least we had learned some very useful skills and had been inducted into the pleasures of the sewing circle, which even for the very young creates an opportunity for sharing and hearing stories, especially when the class teacher, somewhat unusually, desists from banning conversation.

I still enjoy a good sewing circle and am fortunate that my profession allows me to travel so much around the world and meet kindred spirits from different walks of life, all sharing the common interest of textiles and hand making. Inevitably small talk turns to gentle reminiscence, sometimes even to rounds of song (especially in New Zealand). Through my wanderings I am privileged to hear lots of wonderful stories over cups of tea, when teaching classes or simply while in transit from one point to the next. During one comfortable stitching session a participant shared a story about a friend of hers who was blessed with an abundance of siblings, all sisters. The mother would simply buy a bolt of cloth each year and make every girl a dress of the same colour, fabric and style, with the only variation being the size of the garment. I imagine it made identifying the family in a crowd rather easier.

a stitching sampler worked as a small bag, made by my mother as a child

Another woman told of her Sunday best dress, sewn from a lovely heavy silk and cut several sizes too big so that she was swamped by it when it was first made. The dress had an extra-long hem that was let down a little further every so often until there was no more spare cloth and a false hem had to be created using another piece of fabric. By this time the once-capacious garment was straining at the side seams and doubtless (like the curate's egg) was only good in parts, but these resourceful and economic design features meant one dress could be made to last for up to seven years. The youngest child in a big family might well have had a number of garments to wear but rarely something new of their own.

During the war years of the 1940s there was strict rationing of resources on all sides of the conflict. In Britain dresses were limited to lengths above the knee and such frivolities as pleats and frills absolutely banned. Further frugality was encouraged by the need to exchange certain numbers of coupons (as well as money) for clothes or cloth. Those who are serious about minimising their use of cloth might wish to apply such design rules for themselves.

a seductive selection of undyed natural fibre cloth at one of my favourite fabric merchants in Wellington, New Zealand ▶

Material matters

Choice

When purchasing new materials for garment making it is important to study the content of the fabric very carefully. If you have any doubt at all about whether the label or the salesperson is telling the truth about the presence of synthetics, there are a couple of ways to test this for yourself. The simplest means of testing for synthetics is to attempt to crease a corner of the cloth. If it springs back without a visible wrinkle then it is probably synthetic. Sometimes, however, synthetic fabrics can be creased, so if you are still not quite sure then a burn test is advised. Ask for a swatch of the fabric, take it outside in the fresh air and use a match or cigarette lighter to apply a flame to one end. The smell of burning plastic or evidence of small melted bits and blobs along the charred edge will alert you to unsavoury inclusions. But given that more often than not naked flames are frowned upon in public places, it is useful to have acquired the knack of string-making; attempting to twine string from a thin strip of fabric is a very good test for synthetic content, as only natural fibre fabrics will retain the twist. All of the others should bounce straight back at you.

Choose the best quality you can afford and consider whether you want to support ecologically compromising multi-national dye companies by purchasing synthetically dyed cloth. Many fabrics that have been grown and processed without the application of harmful substances are then still dyed using unnatural colours and for some unfathomable reason still labelled as 'organic'. Even if the label tells you that the dye is safe because it was applied using water as the carrier, unless it is a vegetable-derived colour or a user-friendly mineral such as ochre it still has the potential for harm. The vibrant colours that are so popular in contemporary fashion take a heavy toll on the environment both in the process of making the dyes and in their application. Huge quantities of water and heat energy are used in commercial dyeing. Responsible manufacturers will recycle the water and save heat energy by holding the liquid in insulated underground storage tanks, but given the serious financial commitment required to invest in such environmentally friendly infrastructure such approaches are sadly few and far between. That said, the traditional processes whereby vegetable dyes are applied sometimes also consume excessive water and energy. Using the eco-print technique (where the

substrate water can be continually recycled and topped up) or naturally derived and fermented indigo are two means whereby beautiful and ecologically sustainable colours can be utilised to bring individuality to garments.

If you intend to purchase pre-dyed fabrics it makes good sense to take a swatch home before committing to the final acquisition. Place the swatch in a non-precious receptacle and pour on a little boiling water. Leave it to sit overnight. If the liquid is at all coloured in the morning I would recommend against purchasing the fabric. After all, what can come off in the wash can also be released against your skin – and worse still, be absorbed by it.

a treasure trove of pre-loved Japanese garments provides a rich source of material; unpick and wash before use

Hem, wash and press

Before cutting out your pattern it is imperative to wash your cloth so that accidental shrinkage will not ruin your garment. Cotton and wool both tend to shrink rather more than silk, linen or hemp. Use the laundering process that you plan to apply to the finished garment throughout its life – that way there will be no nasty surprises. Bear in mind, though, that felting is a continuous and irreversible process and that hot washing of woollen items will inevitably make them a little smaller each time. Hemming the cloth before washing will prevent accidental fraying. After laundering and air-drying the piece, press it flat to make it easier to cut the pattern. I've always been a little mystified by the choice of settings on irons, as most natural fibres can actually be pressed with a hot iron. Wool needs far higher temperatures to catch alight than any other fibre, yet the ironing setting is usually one of the lowest. Silk is unreeled from the cocoon in boiling water and it, too, is given a low position on the dial. Perhaps the reason is to save the manufacturer of the iron from potential litigation.

At school we were taught to iron the paper pattern as well. I scoffed at this process at the time but am now happy to endorse pressing patterns, as those that have been jammed into a packet can be extremely wrinkly and therefore difficult to smooth out or pin down. If you have drafted your own pattern on newsprint or other paper then there is less likely to be a problem.

Unless I plan to use the selvedge as a no-sew hem, I remove it before cutting the pattern and save it for making string. I also find the selvedge useful for tying up eco-print[25] bundles, adding to the surface of a textured felt or for stitching onto other garments as a trim. It is best to carefully cut it off with a sharp pair of scissors rather than ripping it, as for some reason a rip made along the warp of a fabric will often wander off to one side.

The matter of patterns

One need not be restricted to traditional paper patterns, or to designing and drawing one's own. Often an excellent pattern can be produced by unpicking a worn-out and beloved garment or by taking apart a garment found in an op-shop. Scan the racks with an eye for shape and cut, trying to look past ghastly polyester fabrics or hideous surface designs. It is surprising how many interesting garments out there are disguised this way; in the end it's a matter of taste and personal preference. I once owned a soft, comfortable and rather lovely cotton blouse that I wore until parts of it were absolutely transparent and ready to crumble. To my eyes, then unfamiliar with the Japanese art of *boro* or the Indian concept of *kantha* (both words that simply mean 'rag'), it was beyond even mending and the style was no longer available in the store, so I carefully unpicked the seams, pressed it and made

my atelier walls

several more in the same design. These days I would undoubtedly layer on the patches. When you find a suitable discarded garment, wash it and either unpick the seams or cut carefully along them. Interesting features or things to make note of as well as the name of the pattern part can be written directly on the cloth using a felt marker pen. Remember to line up the grain of the 'pattern' piece with the grain of the cloth and to add on a seam allowance before you begin to cut. Alternatively, if you have found a treasure that you are really enamoured of and cannot bear to destroy, you can trace each part onto paper, making careful note of all the important details.

Remember, though, that taking apart a design belonging to someone else in order to reproduce it and profit from it is an infringement of copyright and therefore illegal. If you are making garments commercially you will need to devise your own designs and would no doubt be quite upset if someone else were to copy them.

low-waste pattern cutting – following the pattern to open a tube of knitted cloth rather than making a straight cut

the finished pair of hemp-knit yoga pants, based on a pattern cut from a pair of complimentary airline pyjamas that had outlived their usefulness

the yoga pants drying in the sunshine after a soak in milk (in preparation for dyeing)

New or old?

When selecting material for a sewing project it is important to consider the relative strength of the fabrics. New cloth will not pose a problem, particularly if the garment is to be made from one type of cloth only. The challenge begins when we are putting together a garment using textiles from a range of sources. Old should never be used together with new unless the fabrics exhibit a similar tensile strength. You can test this in a very rudimentary fashion by joining two dissimilar pieces with a line of machine stitching and then spending some time (perhaps while watching a sunset or being a passenger in a car) gently tugging at both ends of the sample. If they are of different strengths you should soon have the evidence before your eyes, as the stronger one will pull holes in the weaker one.

What to do? When juxtaposing old materials such as vintage kimono silks with something sturdier, I reinforce the silk from behind using something such as a soft repurposed cotton sheet, or parts of old pillowcases. A few lines of running stitch through the relevant pieces will help to distribute the pulling forces. In general, though, it is best to observe the rule of matching fabric strengths.

kimono have been designed with future unpicking in mind and come apart into neat lengths of cloth

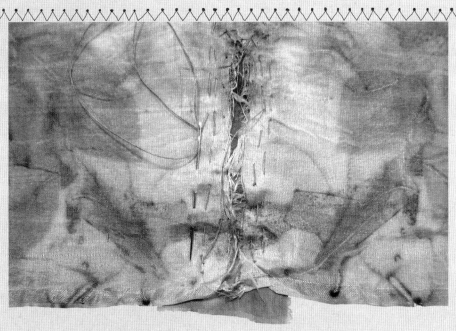

simple mending with running stitch

Bits and pieces

What equipment do we need to make clothes? The choice of machines and overlockers and complicated gadgets is almost overwhelming and while I've been guilty myself of seduction by object (such as the overlocker acquired to assist in the construction of dance costumes some years back), my favourite means of sewing is still the simple use of a hand-held needle and thread. It is also (for me) the least likely to end in accidental disaster, as the slow and gentle method of working allows one to correct mistakes before they grow too big.

A couple of years ago I accepted the commission to design and make a wedding dress. It was a little daunting, as the client and I had never met and I had to rely on photographic evidence as well as placing my trust in the accuracy of her measurements to get an idea of her general appearance and size so as to be able to create a shape that would suit as well as one that actually fitted. The bride sent me a wonderful box of family treasures that were to be incorporated in the dress, including lace-edged doilies and tiny children's garments. We agreed that the underlying structure of the dress should be a heavy silk satin and to my delight I found just the right fabric at my favourite silk supplier. It had been very slightly water-damaged in transit and consequently had an interesting distribution of marks to provide a cloudy shimmer to the background. It was clear from the start that the whole garment would have to be hand stitched with a much more robust thread than could be fed through the sewing machine. This way the seams would be more resilient and it would also be easier to affix and insert the various precious pieces that the client wished to have included. The pieces were of quite different provenance – some silk, some linen and some cotton

the wedding dress, with fragments of children's clothing visible as part of the 'train'

– so the embroidery thread served also to harmonise the whole, the *kantha* stitches blurring the boundaries of the different fabrics. Being rather slower, hand sewing also allows mishaps to be spotted and rectified before they become too dramatic. Better still, the slow approach means that disasters occasioned by speed-sewing with machines tend not to happen at all.

▲ the dress being rinsed of leaf matter in the river

◄ the bride unrolling the dress after its baptism in the dye-pot

the first fitting of the wedding dress; doilies and ► lace fragments adorn the surface

On completion of the garment, I travelled to meet the client and must confess to a few private tears of relief when she put the dress on for the first time and it fitted like a glove. The looming thought of possibly unpicking the hand stitching to make adjustments had hung over my head like a small but persistent storm cloud and it was a pleasant feeling to have it waft away.

We completed the work on the dress together, taking a 'windfall walk' to collect leaves in the forest where the ceremony was to be held, bundling and cooking the dress in a cauldron over an open fire then taking it back to the forest for a stone-secured benedictory rinse in the river that flowed there.

No adjunct chemical mordants were used and the leaves had been removed, so we felt it unlikely that the symbolic immersion of the garment would harm the waterway. It is this kind of slow and mindful journey of making that I find so very satisfying, the quiet hours of gentle stitching followed by the vigour (and possible surprise) of the flame-fed dye cauldron.

Occasionally the finding of spectacular plant material has triggered the making of a garment. Visiting the Victorian town of Jamieson, I stuffed my apron pockets with windfall leaves from the beautiful Japanese maples there (*Acer palmatum*) as well as harvesting flowers from St John's wort (*Hypericum perforatum*). The latter is a noxious weed in Australia and picking it for the dye-pot helps to reduce the potential for the plant to spread by seed. Later that evening, inspired by my botanical collection and wanting to make a useful souvenir of my excursion, I hand stitched a simple shift dress (pictured on page 96) using a lovely piece of jacquard silk that I had been given by a friend. The hand sewing of the back seam, neck and sleeve edges took only a couple of hours; a pleasant occupation while musing over the doings of the day. Once the garment was completed I laid it on the floor, spread my botanical bounty over it and bundled it around a lovely stone that I had borrowed from the Jamieson River[26] before boiling it for a while. I let it cool in the dye-pot overnight and the next morning I had a delightful present to open. I treasure the dress and the memory of plants and place imprinted on the silk, a practical memento of a lovely day and evening.

the imprint of Japanese maple leaves (*Acer palmatum*) on silk ▶

While I always sew by hand during my travels, the various helpful gadgets in my sewing room include the aforementioned overlocker, a 1965 Elna electric sewing machine passed to me by my mother and the wonderful hand-cranked 1927 Singer sewing machine that my grandmother lugged through war-torn Europe. The latter is very much my favourite mechanical device. Many hand-powered sewing machines were later modified to take advantage of electricity; they are relatively rare in their original form.

Other necessaries include a number of tape measures (at least one is always lurking somewhere under a pile of cloth), pearl-headed pins to keep the various bits together, a large collection of scissors (as much because they're beautiful to look at as well as vital for cutting) and some tailor's chalk for marking cloth. I also keep a small container of ochre pieces that are used as a chalk substitute to make marks, especially if I wish the marks to stay longer in the cloth. (Ochre withstands over-dyeing and laundering remarkably well.)

A big roll of newsprint paper is always handy for drafting patterns and a box of thrift-store cotton sheets offers the perfect material for sewing toiles. Old sheeting is wonderfully soft and comfortable, so toiles that have been made for a commission are eventually repurposed to become useful garments in their own right. Very little is wasted and even the tiniest scrap finds a home on a larger work sooner or later.

The various machines I operate are always loaded with cream-coloured silk/cotton thread so that the seams will eventually pick up colour as much as the rest of the cloth. Everything I make begins white, greige (the industry appellation for the colour of raw or undyed cloth) or sun-bleached and is only coloured (with plant dyes) after the final seam is sewn.

A tailor's dummy or dress-form is an invaluable piece of equipment. Being able to pin and baste pieces of cloth onto a three-dimensional object that is of a similar size to your body is so much easier than having to draft precise patterns. Nipping and tucking as you wander along the cloth, bunching and trimming as desired, turns dressmaking into a delightful adventure.

Putting your own stamp on things

The decision to put the fickle world of fashion behind you and commit to a wardrobe that has less impact on the environment doesn't necessarily mean a lifetime spent wearing hessian sacks tied up with string and shoes crocheted from hemp. Instead it provides impetus for developing a signature look and can make the whole process of managing a wardrobe so much easier. Finding a style and deciding on a silhouette that suits your body shape can be an entertaining activity (see Chapter 8). Having generally fewer clothes and choosing or making colour-related garments that can mix and match makes for a simpler life. It also makes for substantial savings in the budget.

I have a couple of patterns that I simply reproduce with minor variations as required, including trousers, skirts, shift dresses, aprons and cardigans. Packing for travel becomes a breeze as the various components can be interchanged for a different look every day. An immersion in a plant-

derived dye-bath makes each piece unique but still related to all the other pieces. As things become worn they acquire patches and repairs; when they become too frail to wear they are repurposed. Most of my shirts are found at opportunity shops and thrift stores but socks and undergarments are still purchased new. Nobody is perfect – I'm a little squeamish, and there are certain clothing items that should, for simple reasons of hygiene, remain personal. In our family we still giggle over the elderly friend who offered to *lend* my grandfather a pair of underpants after he had (along with the rest of us) lost his clothes in a bushfire. Those underpants were already well worn and patched; the prospective lender had been through two world wars, been a displaced person and spent time as a refugee, so his perspective on their value was somewhat coloured by experience.

While wearing clothes that don't differ vastly from one year to the next could be seen as a kind of uniform, this is not necessarily the case. The variations offered by the addition of scarves, beads or fresh flowers are almost endless. I seem to have developed a style that essentially blends very basic eastern European and central Asian folk costume with a few more contemporary pieces; it has been a subconscious return to the cultural roots of my ancestors that simply evolved naturally from making comfort and practicality the dominant deciding factors in clothing choice. Wearing trousers under skirts makes it easy to climb, ride or sit without having to worry about modesty. Wearing an apron provides me with vital pocket space as well as something to dry my hands on when the need arises. The layers of cardigans and shawls are rather easier to adjust to the weather than a bulky jumper and my big coat doubles as a blanket when I am travelling. Sturdy pull-on boots protect my feet from snakes, embers and inadvertent splashes of hot water as well as being easy to pull on and off when entering temples, homes and other sacred spaces. Drawstring trousers can be slept in comfortably and my wool socks are cool in summer and warm in winter. Everything in the wardrobe goes with everything else as all of it has been duly baptised in a pot of leaf-water at least once. I can dress in the dark and know that, providing the layers are

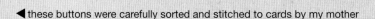
◀ these buttons were carefully sorted and stitched to cards by my mother

Jo Kinross (New Zealand) hand-knitted merino socks that are comfortable all year round

on in the right order, I can step out the door with confidence. Admittedly my voluminous clothing does cause raised eyebrows at international airports where I am invariably subjected to the double-strength frisk in case I am concealing a small family of terrorists under my skirts.

The thrill of the chase

Hunting and gathering materials to create your own beautiful wardrobe is, of course, half the fun. I have been known to frequent thrift stores and opportunity shops with great fervour. Flea markets, second-hand emporia, garage and clearing sales are other good sources of pre-loved materials. Even dump-ground sales outlets often have a surprising quantity of excellent clothes and textiles available – I am frequently astonished by what people will throw away! I find great pleasure in carefully unpicking an old stained kimono or in pondering how I might transform a couple of wool blankets into a solid winter coat. At the fabric stores I find on my travels I first give my attention to the remnant box, where unused or slightly marked treasure can often be found at a price somewhat reduced from that on the roll.

Auction houses sometimes offer bolts of fabric for sale; some from drapers that might have gone out of business, and others that might have been water damaged during floods or smoke damaged as a result of fire. Providing the cloth isn't full of holes it has much potential, and often can be bought quite cheaply.

The search is not merely limited to cloth, either. Buttons are a constant source of delight and the hunt for interesting ones can easily become an obsession. I've seen a couple of avid seekers nearly come to blows over the button box in a charity shop. Precious old buttons made of jet, glass, crystal or pearl shell can elevate a garment from mundane to unique, and who knows where they have been or what stories they could tell? I will sometimes purchase old stained garments if the buttons are good enough, remove the treasure and then if the cloth is unsuitable for my purposes or cannot be made into string, deliver the remains to the animal welfare shelter for use as bedding. At least that way there's relatively little wastage.

Pre-used patterns are also worth looking for. If you are really lucky you might find gorgeous vintage styles that have long been out of print. Patterns that don't match your body size can be adjusted or simply used as inspiration. With a little ingenuity pattern pieces from different designs can be mixed and matched to create unique combinations. My mother frequently used pieces from different patterns in the quest for interesting details that would stand out from the crowd. She had a splendid collection of patterns dating back to the 1950s; even idly shuffling through such things can be quite inspiring.

a pair of vintage Japanese hemp trousers, showing the careful reinforcing of the seams. The trousers have a button fly and have great potential for recommissioning as a comfortable skirt

If purchasing pre-loved zippers, remember to check them carefully for missing teeth. Few things in dressmaking are more dispiriting than having to unpick and replace a carefully inserted and firmly stitched-in zipper. Taking things apart with the intent of using them for something else can be enjoyable; unpicking your own laborious work and feeling that it has been a waste of time is no fun at all.

chapter 6
Maintenance

Perhaps we should never procure a new suit, however
ragged or dirty the old, until we have so conducted,
so enterprised or sailed in some way, that we feel like
new men in the old, and that to retain it would be
like keeping new wine in old bottles.

Henry David Thoreau [27]

There's no arguing with the fact that clean clothes are important. Nobody likes to sit next to you on the bus if your clothes smell and most of us have at some time experienced the discomfort that comes with wearing something just a little too long, whether on a long-distance flight or as a result of trekking in the mountains for a while. In times past, having clean clothes either showed that you had applied yourself to get them into this state, or could afford to hire someone else to do so for you.

My grandmother used to say that if your clothes were washed, pressed and mended you could hold your head up and go anywhere with dignity. Of course when I was a teenager I sniffed at this notion – who would be seen dead in something that was conservative, no matter how clean or mended?

The tips and tricks first published by the British Ministry of Information in 1943 (as part of wartime austerity measures) read for the most part as a guide to ecologically sustainable living, if we utilise bio-friendly cleaning agents. Their suggestion for softening 'hard' water by adding a little ammonia could be carried out by the use of fermented urine (an excellent source of ammonia) if you were really keen. I have found that a couple of rhubarb leaves simmered in water and then strained into the wash-bath (or dye-pot, for that matter) works wonders; although the chemistry of it has me baffled, as ammonia is alkaline and the rhubarb brew is acidic.

Sensible washing instructions

Wash clothes less often, at lower temperatures and using eco-detergents, hang-dry them and avoid ironing where possible.[28]

dye samples line-drying in the shade of a Belgian garden – air-drying your clothes on a washing line in a similar manner will have them smelling fresh and sweet

A little care

Making clothes last longer is largely a matter of common sense and taking a little more time over detail. It makes sense to try to reduce the number of times a garment is pounded by a washing machine or bleached in the sun and to keep in mind that old but apt proverb 'A stitch in time saves nine'.

- mend clothes before washing so that holes and rips don't grow
- wear old clothes at home
- air clothes before storing, either indoors or outside in the shade
- hang up clothes while they are still warm (they'll be less likely to crease and wrinkle)
- remove stains while they are fresh
- fix that loose button before it wanders off into the world
- rotate your clothes and shoes; they'll wear better
- don't let clothes become really filthy; encrusted and ingrained dirt is difficult to remove
- save energy when washing by soaking clothes (unless the colour is likely to run)
- sponge clean the surfaces of coats and jackets with a damp lint-free cloth before putting away

From Grandfather's dressing table

long- and short-bristled brushes to remove fluff from coats together with a white chalk cover stick intended to keep shirts looking spotless

When things do need cleaning, take a little time to think about the best way to treat the garment. Hand washing where possible will reap huge benefits in extending the life of a favourite.

Make do and mend – cleaning methods

Someone once told me that during the war years of the 1940s it was recommended that people rub the outside of their leather shoes with the inside part of a banana peel then buff to a shine with a soft, lint-free cloth. This actually works rather well, although I do wonder whether the British ever saw a banana during those straitened times of strict rationing. It's a common fruit these days, though, and if you don't have inhibitions about purchasing food from outside your own bio-region (or if you live in the tropics) then it's a fine way to shine your shoes. Bury the leftover skin in your vegetable patch or under a rose bush, as it's rich in potassium and an excellent slow-release fertiliser.

Those who keep sheep or have farming friends will have another useful source of polishing aids. Greasy wool straight from the sheep is rich in lanolin and perfect for shining leather shoes. Anyone who has helped at shearing time by jumping into the wool press to squash down the fleeces will attest to the reward of beautifully polished work boots.

airing bedding in Yamaguchi, Japan

Using the world outside

Snow can be a helpful cleaning agent for those in wintry climes. To freshen jumpers and sweaters simply beat them (in their dry state) against a nice firm snow bank – not powder or wet snow because that will simply stick to the garment. On a good solid snowdrift the garment will lose quite a lot of filth and begin to look much brighter, leaving a grubby patch of snow as evidence of the activity.

This is also a splendid way to clean small rugs and carpets. If the temperatures are low enough in your region you could even leave the carpet outside overnight to freeze. This will inhibit the bacteria that reside within and help keep the carpet smelling sweet. It is very important to let it thaw before trying to roll it up and carry it indoors, as frozen cloth has been known to simply snap in two, which might be considered undesirable.

Some denim producers now suggest that the eco-friendly way to manage your jeans is simply to freeze them once a week to kill bacteria and stop them smelling. Bearing in mind that bacteria have survived over thousands of years in permafrost, I suspect this may be wishful thinking, but (as with the carpets) extreme cold will certainly inhibit their growth. Although freezing rather than washing will maintain your jeans in good colour longer than regular mechanical washing, it won't get rid of stains. So either (if you're the grubby type) you can watch your jeans gradually collect a series of marks from various life activities, or you could combine the two methods and occasionally

knitwear can be cleaned and refreshed by beating it against a dry snow bank

immerse them in warm soapy water to clean the surface and in between times pop them in the freezer to stop them becoming too whiffy. Remember, though, that no amount of freezing is going to take away the aroma of a long night spent in the company of smokers in a public house and the jeans will need to be thawed before wearing to avoid accidental 'breakage' of the fabric. On the other hand, the freezer could become an intriguing 'distressing tool' for those keen on customising their denims.

Summer carpet washing

During my teenage years we lived in what was originally conceived as a 'summer' house on the ridge of Mount Lofty. In those days the area had not yet been built out by new housing and was considered by the residents of Adelaide to be a destination that could only be achieved with the accompaniment of a compass and a cut lunch. When my parents attended dinner parties in 'town', others at the dinner table would express astonishment that they were actually planning to drive back up into the hills at night. This relative isolation meant that the residents along the ridge formed a friendly community and so it was that I scored one of my more amusing summer jobs. A little further up our road there was a most magical house called 'Arthur's Seat', a two-storey residence with a smallish tower and about thirty rooms inhabited by an equally extraordinary person. Nancy Harford, who was then, I was given to understand, in her sixties, had buried at least two husbands and was worshipped by a coterie of stylish older gentlemen. She was still overseeing the restoration of her house, a task that had occupied her since the Army had requisitioned (and trashed) it during the 1940s. This was a challenge she took very seriously, travelling to Paris with fragments of wallpaper in order to have the same pattern reproduced by the original company and deliberating over the exact shade of ochre red to make the most of the afternoon light on the walls in the study.

Nan had the most amazing collection of clothes, mostly handmade especially for her, and she wore beautifully cut coats and suits, some of them more than thirty years old. Few people these days will maintain a consistent size over several decades; Mrs Harford smoked, religiously honoured the cocktail hour, worked like a beaver in her house and garden, had legs like Marlene Dietrich's and her dress size never varied from a standard size 12. My summer job was to help drag her magnificent collection of Persian carpets out onto the terrace, give them a good beating and then to wash them using a bar of Velvet laundry soap, a broom and rainwater from the garden hose. There's something very satisfying about sliding around on a soapy Persian carpet with bare feet. Once rinsed and sun-dried the carpets were restored to the various rooms. Even though her house was relatively close to ours I'd sometimes be invited to stay for the night at the end of such a day, then be treated to a gin and tonic, dinner and a viewing of the treasures in Nan's wardrobe together with stories about her childhood in the Mount Crawford forest. Later I'd be luxuriously tucked into a four-poster bed with a pile of old fashion magazines and a late-night cup of tea.
The house is a ruin now and Nancy Harford is long buried. But the memories of summer carpet washing still bring a smile and the knowledge that my children are quite right … I really did grow up in 'the good old days'.

'Dry' cleaning

I recommend avoiding dry cleaning at all costs. Many manufacturers will slap a 'dry clean only' label on items for the simple reason that they have no faith in a client's ability to read and carry out washing instructions or because their product is so constructed that they fear normal laundering will reduce its life. When a garment comprises a number of materials that are likely to behave differently when faced with water and a wetting agent, manufacturers take the easy way out and suggest dry cleaning. The truth of the matter is that almost anything can be washed by hand if there is sufficient understanding of how the fibres are likely to react. The very appellation 'dry cleaning' is in fact a complete misnomer – no water is used, but the process is hardly dry as it certainly involves liquids.

The process was discovered in the mid-1800s by a Frenchman, Jean Baptiste Jolly, who observed that his tablecloth was rather cleaner after kerosene from a lamp had been spilled on it. He put two and two together and began to offer a water-free cleaning process, which must have caused a bit of inadvertent excitement from time to time given that gasoline and kerosene were the first solvents used and they are both rather flammable.

manufacturer's insurance against unskilled washing

Garments delivered to professional dry-cleaning establishments are placed in a basket inside a drum filled with a chemical solvent and swilled about in the fluid. The dirty solvent is then extracted by spinning and the garments are 'rinsed' in a fresh lot of fluid. Eventually this too is drained from the machine and the garments spun so that most of the residual liquid is removed by centrifugal force and stored for recycling. After spinning, the garments are gently tossed in a stream of warm air to evaporate all traces of the chemical.

Unfortunately the first solvent bath may already be a recycled one, in which case your beloved party frock might be sloshing about in fluid that already contains nasties from another load of clothes; also, garments are not placed individually in the cleaning machine but are included in groups. The thought of having my clothes tossed about in a carcinogenic petrochemical solvent together with those belonging to complete strangers is frankly disgusting.

Certainly it doesn't seem to be a hygienic practice, more like redistributing collective grime and bacteria evenly over the entire batch of clothing.

Manufacturers recommend dry cleaning simply to protect garments from being thrown into washing machines set on vigorous washing programs. Modern washing machines are easy to use and very convenient but can take their toll on the potential longevity of clothes.

Hand washing

Hand washing delicate garments is so much cleaner and simpler than having them sloshed about in solvents. They'll certainly smell a lot nicer too. Hand-washed clothing will last much longer than that which has been churned in a washing machine. Remember that silk and wool are composed of elements that are very similar to human hair. Both are protein fibres; silk is mostly sericin while wool is composed of keratin (as is human hair). Treat these fibres with the same respect you would give your own hair and they will keep their condition for much longer than they will if sent to the dry-cleaners regularly.

While hand washing can rightly be regarded as excessive toil at times, there is a considerable difference between lovingly laundering your favourite dress (and being grateful for having such a beautiful object) and having to sit sideways on the bath doing the entire family wash; something my mother actually did for the first seven years of her marriage. I too confess to harbouring serious feelings of resentment at having to wash clothes in a large plastic garbage bin while living off the power grid in Andamooka in the far north of South Australia. Washing one's own clothes and those of one's own small children is no problem, but being expected to wash for other adults can be dispiriting. I rebelled when a visiting relative casually slung her underpants into the soaking bucket. These days I hand wash my treasures and use an energy-efficient front-loading washing machine to take care of the rest. I dry sheets, towels, shirts and jeans in the sun and everything else under cover in the shade of the veranda. In winter we use a drying loft above the kitchen.

There's no reason hand laundering garments for oneself shouldn't be a pleasurable activity. After all, the careful washing of precious things honours the object and its maker through respectful treatment given to the garment; it will of course extend the life of the garment as well. Choose a pH-neutral wetting agent that is kind to your skin and that has a fragrance that you like. It doesn't necessarily need to be laundry detergent. I find that a drop or two of shampoo is ideal for washing both wool and silk; these are protein fibres and display properties comparable to those of to human hair.

Alternatively, choose some nice bath soap and swill it about in the washing water to dissolve. It takes very little to break the surface tension and help remove the grime from clothes. Soap has a slight advantage over detergent in that although it might form an unpleasant-looking sludge in waterways (should it get that far), at least that sludge does less harm to amphibians

clothes being washed in a creek in India, along with a local buffalo

line-drying a dress and scarf; above right: mending worn spots on a linen shirt

living in the water. Detergent, on the other hand, stays in solution and (if present in sufficient concentration) can cause frogs to drown by reducing their skin's ability to resist water.

Have the temperature of the water-bath lukewarm to bloodheat (about 36°C, or 97°F), as cold washing isn't much fun and can lead to chapped hands. Wool can be washed in warm water provided that the temperature of washing and rinsing waters doesn't vary by more than 5°C (9°F).

If the garment is very dirty, immerse it in the water-bath and let it soak for a couple of hours. Soaking will save scrubbing, which in turn will save wear and tear on the garment. Use the spin cycle of the washing machine to extract the dirty soaking water from the garment before proceeding with the wash.

If the garment is made from wool, the best way to wash it is to lay it flat in the trough, basin or bath and gently press it into the water. Never rub the cloth, as the action will cause felting. Just gently press and squeeze. Let the water drain away from the garment by removing the plug rather than lifting the garment out of the water. Wool can hold up to 30 per cent of its own weight in water before it even feels damp, so the weight of water held in a wet garment can cause serious stretching (information that can of course be used to advantage if need be).

Once the water has drained from the garment, squeeze it gently to expel more moisture, on no account wringing the item. If you have confidence in the resilience of your woollen piece, use the spin cycle of the washing machine to extract the last drips, otherwise roll it in a towel, place the roll on the floor and walk about on it. This will force more water out by evenly distributing the water content of the bundle between the garment and the towel. Unroll the sodden mass and lay the item out on a fresh towel, gently pulling it into the shape you would like it to be. Leave to dry, turning after a few hours.

Silk can be treated a little more vigorously than wool, but as mentioned above, wringing of either fibre should be avoided at all costs. I never use commercial fabric conditioners, preferring to add a little vinegar to the final rinse for silk and wool to neutralise any residual alkalinity. Fabric conditioners tend to coat the cloth so thoroughly that several washes are needed to remove them from the surface.

Use the spin cycle of the washing machine to extract moisture between soaking, washing and rinsing.

Machine washing

For filthy work clothes and big items such as sheets, a washing machine is much the best answer. An efficient front loader will get things clean using much the same principle as old-fashioned washing by the riverside, when clothes and household linen were wet, soaped and then thumped firmly on the rocks. The machine will use less water and spin the filthy wash water away efficiently so that less rinsing water is required. The washing machine I use takes longer than hand washing but has been designed to allow time to assist in the dissolving of stains by designing periods of soaking into the cycles. Generally I only use about 25 per cent of the quantity of washing powder recommended by the detergent manufacturer, still with excellent outcomes.

Those with limited time at their disposal may need to use washing machines more often than others. Common sense will dictate the sorting of laundry;

coloureds (bitter experience suggests that synthetically dyed reds should be separated from all other washing) kept away from whites, and delicates kept apart from the mud-plastered trousers worn in the garden. While it's good to maximise energy use by filling the machine when possible, over-filling simply results in an even distribution of filth among the contents. Again, good sense will ensure that smaller loads are washed with lower water levels and adjusted quantities of soap or detergent.

Washing on the run

When travelling I simply dump grubby clothes in the shower and let the water from washing and rinsing my hair begin to do the work for me. Stamping on the items helps clean the feet as well as loosening any grime. This is followed by a quick rinse in the hotel sink, after which I press out any excess water using the supplied towels. The lightweight wool and silk garments that make up most of my travel kit dry quite quickly and giving them a good shake out before hanging them up means they'll be fine to wear without ironing. In any event my garments travel scrunched into small silk bags, so crumpling is inevitable, even welcomed.

Fresh air

Frequent airing will help prolong the life of your clothes, allowing smells to dissipate as well as discouraging insect attack by killing moth larvae. Ultraviolet radiation contained in sunlight helps to kill bacteria as well as helpfully bleaching white garments. Unfortunately sunlight does not discriminate between white and coloured items so it is always good practice to hang the latter in the shade. A black T-shirt will be partially grey after a day exposed to strong sunshine. This happens when a garment is being worn as well, but unless one has spent the whole day motionless outdoors it is less likely to be discernible over a short period of time. Compare a new black T-shirt with an old one and the difference in colour should be substantial.

◄ lavender keeps clothes smelling sweet but will not necessarily deter the ravaging moth

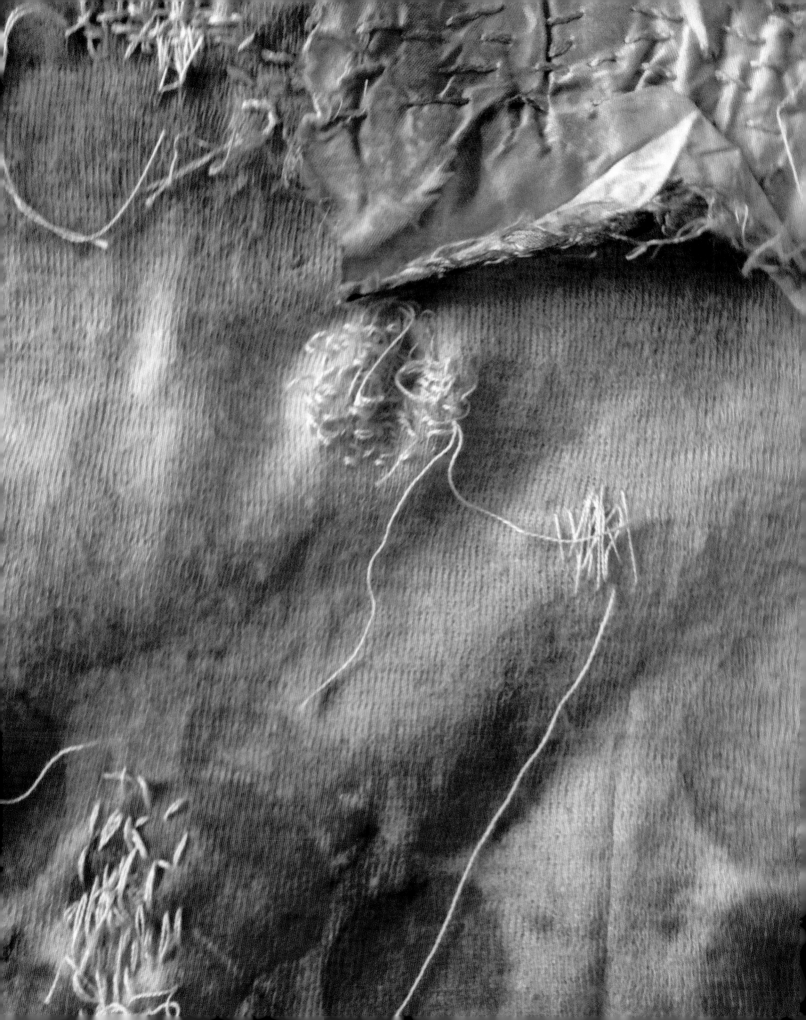

Washing silk

Gently hand wash silk in lukewarm water using a neutral detergent or gentle soap (shampoo is fine). Rinse the garment twice, adding a teaspoon of vinegar to the second rinse. This acts as a conditioner and restores pH balance, as the acid of the vinegar neutralises any alkalinity from the detergent or soap so the fabric feels instantly softer. Excess water can be spun out using the spin cycle of a washing machine, but never scrub or wring silk as this damages the fibres. Remember to dry silk in the shade; sunshine has a deleterious effect on wet silk, causing it to become brittle.

Washing wool

Wool can be washed in a similar manner to silk but with the added rule that the temperature of the washing and rinsing baths is not varied by more than 5°C (or 9°F). Temperature variations affect the natural crimp in the individual fibres that make up wool yarns and fabrics and can cause the garment to shrink. Whether the variation is from hot to cold or the other way around, the effect will be the same. Try to keep the movement of the garment to a minimum during the wash, bearing in mind that each little fibre will be trying to wriggle along just like a fish each time it is jiggled. Even wool coats can be washed provided they are simply laid flat in the bath or (preferably) on a piece of screen mesh and hosed rather than moved about in water.

Remember to dry wool flat or in the shape you want it to remain. Wool has a memory for process and removing even small marks left by clothes pegs can be quite difficult. If not dried flat, the weight of the damp fabric is likely to cause the garment to stretch, especially as the moisture will tend to collect at the bottom edge of the item before dripping away, pulling it down.

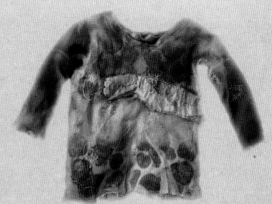

◀ left (detail) and right: woollen baby jumper mended and over-dyed

Stain removal

In general, the best treatment for stains is to catch them as early as possible, flush them thoroughly with clean, cool water and then press the affected part between two thick pieces of cloth or absorbent paper. If the stain is greasy then a little dishwashing detergent in some lukewarm water can be helpful.

If stains are not properly removed and rinsed they can often show up later – for example a lemonade stain might dry clear but will show up as a brown stain when ironing, as any residual sugars caught in the cloth will caramelise with heat.

Red wine

Sprinkling red wine spills with salt doesn't make any sense at all as salt is a mordant and will actually help to fix colour from the wine in the fabric. Flushing the spill with soda water before it has had a chance to dry will help to prevent the wine from creating a permanent mark. But even if it seems invisible after treatment it is very likely to pop up again, making triumphant 'huzzah' noises, if the cloth is ever immersed in some sort of dye-bath.

Blood

Bloodstains can be tricky to remove. If the stain is small and is your own, try wetting it well with saliva. The enzymes in the saliva help to loosen the proteins in the blood from their bonds with the cloth. If there's a lot of blood and it is fresh, take the item outside and hose it well with cold water. Then soak it in an enzyme-based soaking agent before washing. Unfortunately that enzyme-rich stain exploder may also affect colour, whether synthetic or plant derived. The good thing is that plant-dyed garments can be refreshed in the next available dye-pot using the bloodstain as a mordant!

Rust

Rubbing lemon juice into the stain and then sprinkling it with a little salt may quite easily remove rust stains. Alternatively take the garment to the sea for a good soak and then apply the lemon juice.

Grease

The old-fashioned remedy for removing grease, popular before the invention of detergents, involved pressing the affected part between two pieces of clean blotting paper. This might remove excess fatty material from the surface but will also have the effect of sending the remnant of the stain more deeply into the fibres that form the cloth. A small application of dishwashing detergent using a fingertip and then letting the item rest for half an hour before commencing washing seems much the best way to tackle the problem. In the event that the stain cannot be removed, consider making it a feature by embroidering over the top.

Cleaning found fabrics before storing or using

Grubby cloth attracts moths and other marauding micro-fauna, so it is a good plan to ensure everything is as clean as possible before putting it away into storage. In addition, the longer dirt is left in contact the more firmly it tends to bond with the cloth, just as leaving freshly dyed cloth to cure before first washing greatly enhances the resilience of the dye. Frequent airing helps keep undesirables at bay and reminds you of the lovely treasures stored in your stash.

I did love the smell of a laundry dried outside, it was a good fresh smell; and the shirts and the nightgowns flapping in the breeze on a sunny day were like large white birds, or angels rejoicing, although without any heads.[29]

Ironing

Pressing fabric with a hot iron is something I rarely do these days. I revel in the excuse that ironing uses up valuable energy and will contribute to my ecological footprint. The exceptions are when flattening a seam during construction of a garment or when smoothing out a freshly dyed piece of cloth. In the latter case going over it slowly and levelling the crumples from the surface with the iron brings joy and delight, revealing each nuance of the pattern to the gaze. I will admit to the luxury of smoothing freshly laundered sheets from time to time, but rather than wrestling with voluminous pieces in the laundry I lay them on the bed first, tuck in the corners and then iron the middle part where I intend to sleep.

Simple shaking, pegging and folding of clothes can avoid much of the need for ironing. Giving things a good shake and 'snap' before pegging firmly to stretch the cloth is one of my mother's tricks. I always peg shirts at the bottom of each side seam so that the wind does the work of smoothing while they are drying on the washing line. Careful folding of the garments as they are removed from the washing line will keep them in good order as well as making them easier to store.

Another way of avoiding the need to iron is simply to gently twist the garment and knot it into the form of a skein while it is still slightly damp. This will put a lovely pattern of regular wrinkles into it as well as making it very easy to pack for travel.

a flat iron being used by lamplight during a blackout in a village in India

Mending

In frugal households, cloth was valued and not wasted or thrown out.
The clothing of a deceased family member was passed on and used by the next generation.
When a garment showed signs of wear, it was mended.[30]

The process of mending and cherishing clothes must be almost as old as the concept of clothing. By stitching patches over rips, reinforcing cloth or weaving into a gap with thread we are able to prolong the life of our garments, but mending represents more than a purely pragmatic activity. It can be a pursuit that is also representative of keeping things together when times are tough and maintaining standards when one's world might be crumbling; each stitch invested with love as well as frugality. It is also an activity akin to meditation that can take the worker to a quiet place within the whirlwind.

a patched *adire* cloth from Africa

Mended garments were often seen as something to be ashamed of, denoting poverty, and as such have not been treasured as much nor kept to the same extent as ceremonial clothing or even Sunday best. Almost every family has a wedding dress or two stored somewhere but relatively few will have saved grandmother's old mended apron. Many museums have substantial collections of significant garments formerly owned by the well-to-do, but traditionally the practical garments of the working class are more rarely found. This is in part because the clothes of the poor were often worn until they were utterly ragged, or were recut and transformed into something for another family member. Utilitarian garments were there to serve a purpose but too often were seen as reminders of hardship or as evidence of low socio-economic status.

So the mending (or not) of clothes can tell us quite a lot about people and society.

the woollen vestments of St Francis of Assisi show much wear and mending

In the 1960s Chuzaburo Tanaka, a visionary collector of Japanese household goods, set out to salvage the patched and mended clothing of his rural compatriots which at the time was being rapidly overtaken by Western-style clothing. Families were putting their *boro* clothing in the rubbish rather than admit to what they considered to be the shame of possessing repaired clothes. Tanaka-san began to search out these tattered remnants, walking the roads and paths of the snow country in the northern areas of Honshu. Along with garments, futon covers and other treasures he also collected stories and images, bringing together a vital social history. The region was far too cold for the cultivation of cotton so most of the garments were made from hemp grown, harvested and processed by the farmers and their wives. Every scrap of cloth was precious, every loose thread saved. When garments and household items were beyond repair they were shredded and either re-woven as *saki-ori* (a type of cloth using a weft of interwoven strips of sliced fabric together with a fine cotton or hemp warp) or made into string. The string was used as headbands or to tie various things together and eventually burned as a mosquito repellent. Thus the cloth grown from the earth of the region eventually returned to the soil in the form of ash, a fertiliser for crops to come.

The *rafoogar* of India ply an interesting trade; their profession is that (specifically) of darning Kashmiri shawls with minute stitches and invisible mending to hide the ravages of time, in contrast to the graphical gestural quality of the *kantha* stitching widely practised on the subcontinent. Their careful work is considered most successful when it cannot be seen.

In contrast, this thirty-year-old nightie by Vanda Jackson celebrates its holes with bright and cheerful buttonhole stitch. The cloth might be old and worn but it is also deliciously soft. The enhancement of the holes with colourful stitching is quite delightful.

Vanda Jackson (Australia) – multi-coloured darning and mending in a thirty-year-old nightie

Tackling holes in your clothes

A moth-eaten jumper can be restored beautifully with a little imagination. Simple darning is a good beginning and can form a wonderfully textured decoration when allowed to wander happily all over the surface of the garment. Alternatively, cover the moth-munchies with polka dots cut from a compatible fabric, embroider daisies using the moth hole as the centre or hem the moth hole with buttonhole stitch and encrust it with beautiful beads (thus making a feature of it).

You might also cut a lot more holes into the jumper (an all-over pattern of small leaf-shaped cuts works very well) and then give it a warm vigorous wash to felt the cut edges a bit. After the brutal washing, stretch it back into shape and dry it flat, preferably laid on a towel. Wear it over a finer sweater or a long sleeve T-shirt in a contrasting colour to fully appreciate your new lace-making skills.

Try combining the good bits from a couple of well-worn jumpers. For areas that will be under a bit of stress, using a zigzag stitch or an overlocker is recommended; otherwise simple straight stitching is fine.

from left: filling a hole using blanket stitch; mending an L-shaped tear; merino socks hand knitted by Jo Kinross (New Zealand) and darned by Madeleine Munger (Australia); an example of darning on a fragment of an old kimono

- Reinforce weak places or those that are on the sharp end of wearing such as elbows, underarms and knees to ensure sartorial longevity

- Get back into that old habit of carrying a small mending kit, just in case

- To save wear on hems, take a leaf out of Daisy Bates' book and attach a thin strip of leather on the inside

- For invisible darning, unravel threads from hems and seams and use these to darn with

- Decoratively fill a hole by sewing button stitch around the edge and then following the stitches inwards in a spiral pattern

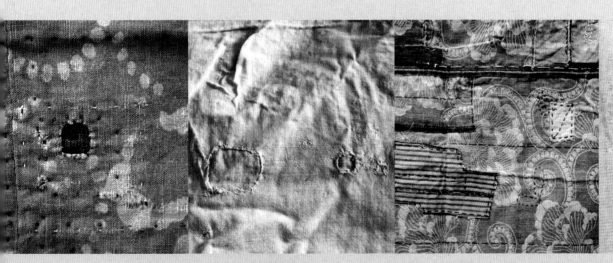

from left: a tiny patch of indigo blue in a *boro* cloth from Japan; mends in an old linen sheet; a delicate mend in a fine wool knit

chapter 7
Gallery

This chapter features the work of clothing designers and artists who have made a practice of working with pre-used or salvaged materials, including some who reference techniques to do with darning and mending in their work.

Emma Christie

Emma Christie (Australia) pursues a way of working that challenges mass production and homogeneity; embracing a holistic approach to design that, while craft-based, looks to more unconventional processes of textile and garment construction. Through stitching, felting and distressing, she hand crafts and customises fabrics, encouraging natural flaws and patterns, endeavouring to create garments that possess mysterious emotional warmth. Combining the raw and unrefined with organic silhouettes, irregular lines and graceful, sinuous forms, she pays homage to the subtle beauty to be found in nature.

Christie's work is influenced by that almost indefinable Japanese aesthetic described as *wabi-sabi*. Her work suggests qualities such as impermanence, humility, asymmetry and imperfection. As a designer and maker she aspires to produce individual exquisitely textured garments that contrast with the impersonal uniformity of convention.

clockwise from main image: ruched metallic skirt, felted belt and top; ruched silk skirt; silk, felt and cotton 'Lichen' dress with pleated plastic bag sash ▶

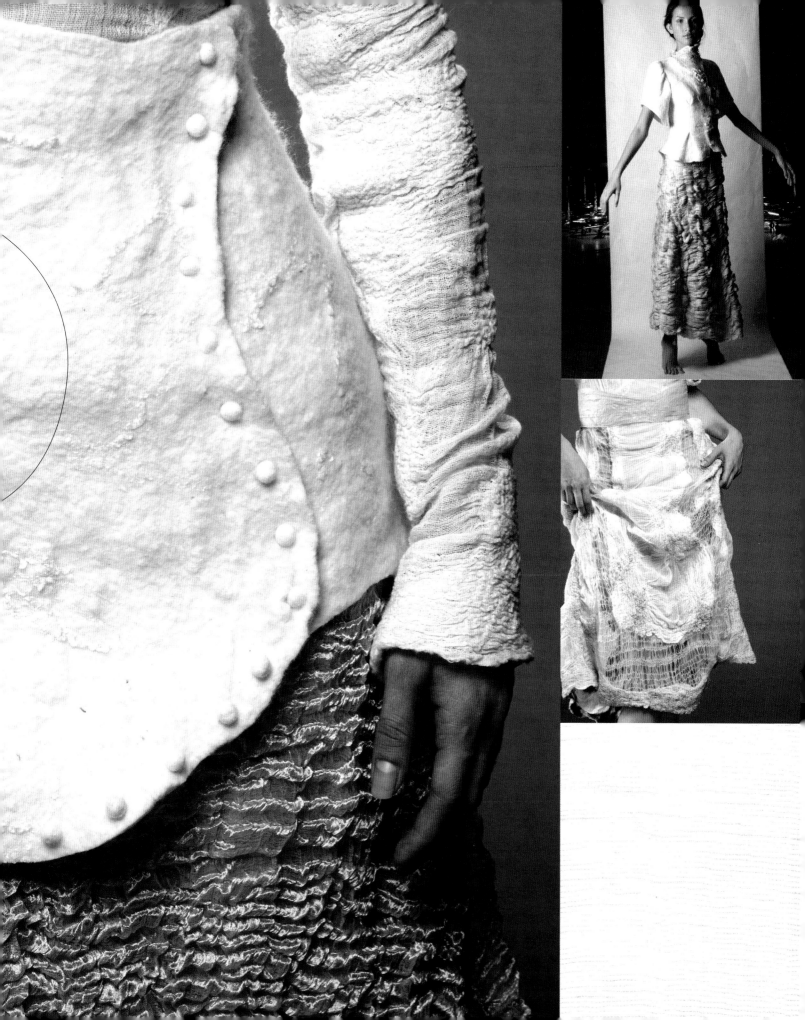

Natalie Chanin

Former costume designer and stylist Natalie Chanin built a company called Project Alabama using a technique based on a form of reverse appliqué, a skill she had learned as a child. The business provided employment for women in the region, many of whom were skilled quilters or had worked in the clothing factories that once flourished in the area (around Florence, Alabama). Initially pre-used repurposed T-shirts were the primary fabric source but as demand for the product grew it outstripped the supply of second-hand goods. Eventually Project Alabama[31] was sold and replaced (in Alabama) by a company called Alabama Chanin, still working with appliqué and embroidery techniques but now using new, organically grown cotton-knit fabric for their output.

The company has recognised that many individuals wish for more involvement in their clothes making and offers an interesting service. Clients can choose one of a number of garment styles from the website and have the cloth cut in the size and colour of their choice and pre-printed with a stitch guide in the pattern of their choice. Alabama Chanin supplies the pre-cut and pre-printed item together with the thread, beads, buttons and other notions needed to finish the work; the client has the pleasure of having been involved in the design of the garment through choosing colour, motif and placement as well as the satisfaction of stitching it all together. There is little waste in the process as the fabric off-cuts remain at the production house for use as embellishments on other pieces and importantly the company is not wasting time, energy and resources in manufacturing huge volumes of garments that might not be sold.

Alabama Chanin 'Bridal' dress

The company supplies ready-made garments as well as custom-made clothing, entirely hand stitched by women in the region and still working with traditional quilting techniques and organic cotton grown in Texas. Because United States law declares that remuneration for piecework performed in the home is illegal (as a means of regulating working conditions), Alabama Chanin has developed an innovative protocol for assigning (and paying for) work done in this way. Workers must demonstrate the skills, care and ability to sew the designs to the standard required by the company through completing a test sample and being accepted into the core group of workers. They are then able to bid on projects, submitting a tender to acquire prepared materials and then stitching the work in their own time at home. In this way the company is able to remain within the law and yet provide useful employment for women living in a depressed rural area that has limited opportunities. The women invest in the work they do by initially purchasing the materials but have substantial motivation to complete the work in good time through the company's guarantee of purchase. At the same time Alabama Chanin is actively supporting and endorsing the hand making of desirable products while maintaining a long-standing regional textile tradition.

Denham

Fashion design company Denham the Jeanmaker (Netherlands) keeps a library of hundreds of vintage and historical garments to use as design references. Rather than simply copying or resurrecting styles they draw conceptual inspiration from the objects, melding sometimes quite disparate cultures and fabrics together with finishing techniques they devise themselves. The results are timeless, eminently wearable and unique, as evidenced in the 'Spy *boro* field jacket', which brings together piecing systems from the traditional mended garments of rural Japan with a shape inspired by a World War II Sahariana field jacket in a tightly tailored model constructed from oiled cotton.

Contrary to popular belief, traditions aren't honoured by blind mindless repetition. Traditions are advanced through systematically challenging convention in a continual search for improvement. We add to tradition through trial-and-error, through innovation and invention. Traditions are ignited and revitalised through inspired acts of respectful revolution.[32]

Denham the Jeanmaker 'Spy *boro* field jacket'

Denham the Jeanmaker 'Spy *boro* field jacket' variations using a range of different fabrics with the same basic pattern

Jude Hill

Jude Hill (USA) works with pre-used cloth, interweaving fragile pieces and linking them with stitches. The 'Magic' cloth is a slow project based on woven components, overlapped and minimally stitched, magically holding together by weaving and mending and put together without any modern-day convenience. It has mostly been made from old recycled scrap and dyed in indigo. It comprises variously linen, silk, hemp and cotton. 'A story cloth base, eventually to become my life's work. It will probably be enormous and never finished and is already in use as a blanket.'

'Shelter' is a new project in progress, in a long format; one of three separate long cloths made from rescued cloth. It includes sections of cloth originally excavated from inside a wall (wallpaper backing), a moth-eaten woollen paisley scarf from Ireland (Grandma), old hand-woven hemp towelling from Turkey and indigo kimono scrap. All fabrics have been compromised by age but are strengthened by weaving them together and therefore given renewed purpose. When attached the three pieces become a blanket; detached, they become shawls or shoulder-warmers.

top: 'Shelter' (detail); bottom: 'Magic' (detail)

Holly Story

Holly Story (Australia) often works with discarded clothing, taking pieces apart and adding stitch and surface decoration.

She brings together materials from indigenous plants with the traditions of domestic needlework: stitching, mending, unpicking, folding, piling and storing while quietly contemplating influences of history and experience. She understands the separateness and dislocation that set apart immigrants and even the children of immigrants from that connection that comes in through the soles of the feet and through the lungs and that courses through the veins.

'Breath' centred on the dispassionate observation of a dress as it lay gradually dissolving on the bed of a river – recording its hastened passing in still and video photography. It might have languished in a cupboard for years; here the artist has accelerated its demise, transforming it into a visual poem that captures the cloth as a butterfly in the net (whose wings are inevitably compromised by such capture). 'Subject to change' was formed through the act of unpicking a linen shirt and allowing the separate pieces to be affected by their placement in the natural environment; Story suggests it might be herself unravelling and being embraced by the landscape.

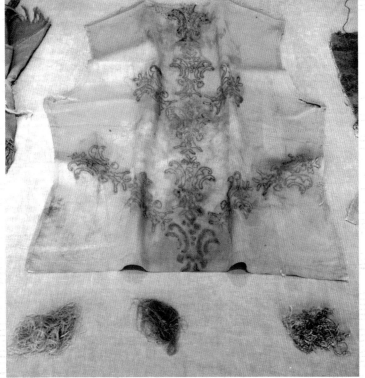

top: 'Breath' video still; bottom: 'Subject to change' (detail) – linen shirt, plant dyes, iron-oxide paint, display case

Christine Mauersberger

Christine Mauersberger (USA) chooses to stitch onto vintage linens and pre-owned bits and pieces of fabric, finding they tend to have a nice hand and good drape; qualities making stitching a pleasurable meditation. She likes to imagine that the gently used materials hold the memory of a previous owner and that in a small way she is involved in their life cycle.

'The marks made by all living things sustain my interest; from the early human markings found on ancient cave walls to the preserved remains of plant and animal life fossilized within the geological record. My marks are based on memories, some are specific, some are general. It's exciting to find vintage linens that still retain the printed blue ink markings intended as a guide for embroidery. Rather than washing them away, I enjoy stitching my own marks over old cloth.'

'Redland II Map'

Roz Hawker

Roz Hawker (Australia) uses rhythmic running stitch to make graphic marks on the surface of fabric; thus homely mending techniques become cloth drawings whose elements are gentle gestures with needle and thread. Deceptively simple at first glance, using a single layer of cloth and a fine thread, the myriad repetitions of this minimal stitch build complex works that invite meditative contemplation. The mending that might be seen by some to be a menial housekeeping task is used here as a means of telling stories about life, narratives that may have slightly different meanings for each viewer.

detail of stitched cloth fragment

Dorothy Caldwell

Dorothy Caldwell (USA) is intrigued by the life of cloth as it breaks down and is repaired, wears out and is reinforced and finally may be reconstructed into something entirely new. Thread-by-thread stitching, mending and repair become cumulative elements that leave a record of lives lived; ordinary everyday cloth encodes time and human history. For Caldwell a simple mended bedsheet is a thing of beauty and a testament to the power of cloth to record the care and work put into those things that touch us in our everyday lives. Caldwell muses, 'The making of my work slows me down and connects me to the place where I live. It reminds me of the "dailyness" of work traditionally carried out by women.'

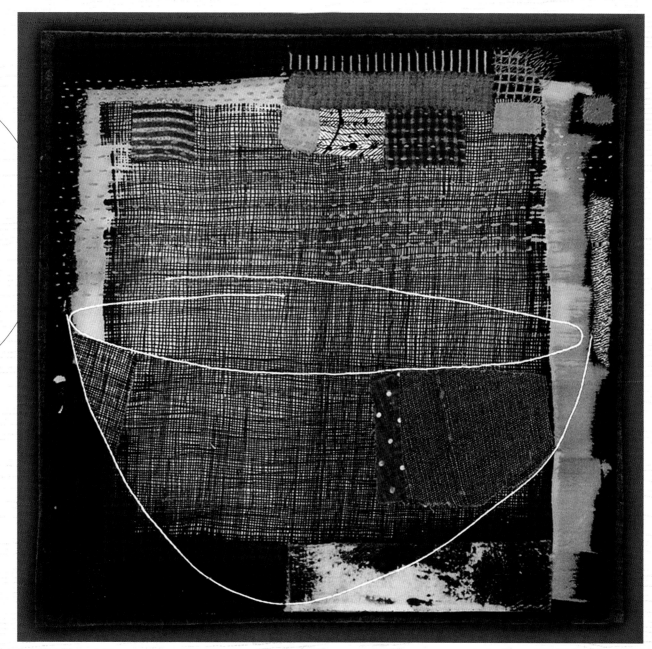

'Lake'

John Parkes

John Parkes (Australia) culls his materials from pre-used utilitarian objects. Sheets, blankets, jeans and pyjamas have all undergone transformation at his hands in a practice that uses stitch and sometimes also plant dyes. Parkes' work with second-hand/recycled/found cloth is underpinned by notions of decomposition and temporality, the recognition that neither cloth nor user will last forever. Whether small fragments to affix talisman-like to clothing or larger pieces for the wall, his works all reflect gentle processes of layering and stitching; the surfaces a textured map created by the wandering path of the needle.

'Genes/Jean's' ▶

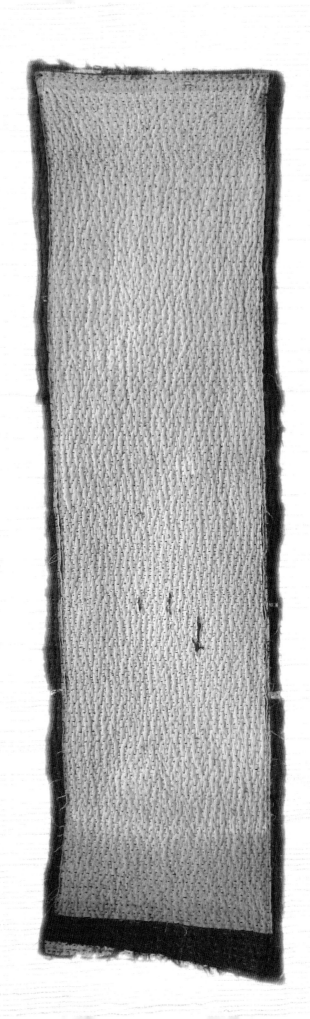

Anu Tuominen

Anu Tuominen (Finland) lives and works in Helsinki. She gathers the lost and the discarded together, sorting them into families and groups, then connecting or arranging them to form objects and installations that are beautiful to look at as well as inviting thought and reflection. She has a way of looking at things from unexpected viewpoints, of exploring possibilities for use beyond the original intent of the producer. It is a form of bricolage, using cutting, stitching, knitting and arranging of familiar objects usually derived from domestic sources, that tells gentle stories while making art from the mundane.

'Shawl'

'Path'

Tracy Willans

Tracy Willans (Australia) bases her work in the premise that earth is sacred and should be nurtured rather than exploited, choosing to work with materials that are ecologically sustainable. Willans works with textiles and painting; experimenting with paints made from kitchen-sourced substances including egg, milk and yoghurt. Her delicately patterned buttons, hand formed from clay and fired using leaves and seaweed are the perfect additions to plant-dyed clothing.

leaves and seaweed colour these hand-made buttons (shown with a small silk bag made by Emma Christie) ▶

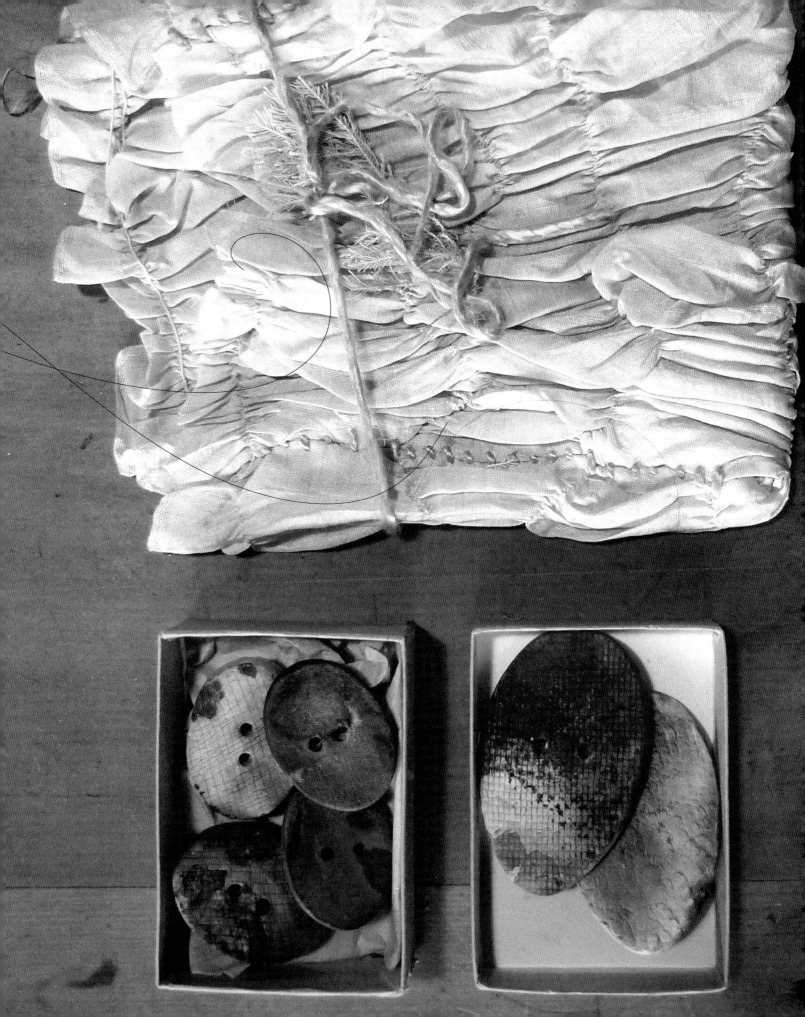

chapter 8
Repurposed and repurposing fashion

Old clothes carry something with them. You can feel the presence of the person who used to wear them.

Mary Lee Bendolph [33]

a simple dress, patched with silks, handstitched and over-dyed with eucalyptus leaves

There comes a time when even the most beloved garments outlive their initial purpose. They might have been outgrown, or have become too large if the owner loses weight. Styles might have changed so much that a once-fashionable garment could look frankly ludicrous when worn in combination with more contemporary things, even though the fabric might still be perfectly good. The excessive shoulder pads of the 1980s are a perfect example of a must-have fashionable detail being regarded as ridiculous only a short while later. If an irremovable stain makes a garment unsuitable for wear, a radical intervention might be called for in order to salvage useful parts. Sometimes only small alterations are needed to refresh a garment and make it a delight to wear again.

It's worth planning for a bit of a wardrobe reshuffle at least once each year and I try (though admittedly not always successfully) to aim for spring and autumn to have a bit of a sort-out so as at least to have my woollies tucked away out of reach of the dreaded moth during the warmer months. Most of my clothing has been reconstructed from thrift-store garments to begin with, but even so I rarely throw anything on a discard pile. Rather than lose an old friend completely I might combine two or three pieces to make a 'new' garment. Throughout the year I patch clothing that is showing signs of wear. An old saying among the frugal rural Japanese farmers of yesteryear – 'Never throw away a piece of cloth big enough to wrap three beans' – is one I have taken very much to heart.

That my wardrobe is based around simple combinations of loose trousers, shift dresses, aprons and soft T-shirts topped off by a cardigan and perhaps a wrap or scarf certainly makes things easy to manage. All of the garments, with the exception of a few pieces made from salvaged black linen and the odd pair of jeans, are coloured using plant dyes, which means there is never any difficulty mixing and matching pieces and multitudes of permutations are possible from a relatively small pool of clothes.

If you haven't had a major wardrobe sorting for a while or wish to revamp your garment collection it can be helpful (and entertaining) to try the following exercise. It's one that students in classes on reconstructed fashion have found to be quite helpful. Have someone take a photo of you while you are clothed in something that fits relatively snugly. Enlarge the print on a photocopier to form an image of yourself that is between 20 and 30 centimetres (8 and 12 inches) tall. Glue the picture onto a piece of cardboard and then trim using a craft knife and cutting mat so that you have a paper doll that is the image of you. This is where the fun begins.

Remember that the camera makes us look slightly chubbier in pictures than we do in life. It sees things through one eye whereas we have the somewhat gentler perspective afforded by binocular vision, so don't panic if the shape you see is not the image of yourself that you have in mind.

Using the wee paper doll you can experiment with shapes and proportions. Let your imagination run wild. Give yourself the freedom to play and create paper clothes with small tabs to hold them in place, working in black and white at first so as not to be distracted by colour.

Once you've settled on a few proportions and shapes it's time to introduce colour and pattern, colouring the forms by hand or making collaged pieces using pages from old magazines. Not only is it an amusing way to spend a rainy afternoon, it is also a great deal less stressful than changing outfits in a department store fitting room where the mirrors are invariably vicious and the lighting makes even the healthiest person look as if they are suffering from some terminal disease.

Use a digital camera to create a library of images and combinations that appeal. You could even put a selection together in a small 'look book' that features your favourite colours, shapes and accessories. The book may be useful in helping to make decisions about what needs culling as well as being a helpful design tool if you decide to de- and re-construct superfluous garments; doing this exercise together with a group of friends and swapping the 'dolls' from time to time can often give new insights into shapes and styles that one mightn't have tried before.

a wine-marked thrift-store silk shirt refreshed with random ruffle attachment and a dip in the dye-pot

chapter 8 Repurposed and repurposing fashion

Sorting out the wardrobe

Clear a space in the bedroom, open the door of the wardrobe and sort the garments into five piles, as listed below. If you have trouble sorting, think about what these clothes might mean to you, why you bought (or made) them in the first place, where you wore them last and whether they evoke certain memories.

Keepers

favourites and necessities that can go back into the wardrobe providing they don't need repairing

Tweakers

'keepers' that need small attentions of some sort, such as the odd repair, replacement of a button or perhaps a refreshing immersion in the dye-bath

Throwers

the stuff that will go to the Salvation Army or the dog basket, or if it's really tatty could be transformed into string or rag rugs

Swappers

things still in good condition that can be taken to a 'swishing', or clothes-swapping party

Rippers

things that don't fit or might be slightly damaged but that have features that you are still fond of. These can be put into the redeployment box either for recutting and sewing new clothes, mending well-worn but beloved items or for making patchwork quilts.

'Recycle, reuse, rethink, repurpose, repair' is a useful mantra that when put into practice can save us a lot of money. Clearly such a philosophy is in strong contrast to what fast-fashion manufacturers would have us do. Recidivist recreational shoppers may take some time to adjust their habits but it's worth persevering.

Over-dyeing clothes can give them a new lease of life. Embroider around any stains first and make them into a decorative feature. Often stains can act as mordants, for example an old red-wine stain often turns into a black splodge when the substrate (garment) is immersed in a eucalyptus dye-bath. Up until World War II the clothes of farm workers in rural Japan were maintained by regular patching and mending followed by a dip in an indigo dye-bath. Not only is indigo a beautiful blue, it also strengthened and reinforced the fibres. Originally derived solely from plants (modern blue jeans are dyed with a synthetic form), indigo must be produced by a fermentation process and is not soluble when in its blue state.

Turn a short skirt into a shopping bag by sewing up the bottom hem and adding a handle. Jeans can be treated in a similar manner, unpicking the seam at the crotch and adjusting it to make it flat before stitching. Alternatively simply stitch across the legs (cutting them off at about mid-thigh) to make a splendid picnic bag – bottles sit well in the legs!

opportunity shops and thrift stores welcome your donations

In the bag

Millions of plastic bags are used around the world every minute, ending up in landfills, along roadsides, in the sea and in the stomachs of unsuspecting animals. Shocked to read of a whale that had literally starved to death because its stomach was filled with throwaway plastic shopping bags, British bag activist Claire Morsman founded a splendid non-profit project. Called Morsbags and built on a strategy of 'sociable guerilla bagging',[34] it encourages people to gather together in small groups she romantically calls 'pods' for bag-sewing circles. Each pod chooses a name and develops a local identity, all the while gathering discarded materials and sewing up shopping bags. These pods then stage 'handouts', visiting local shopping centres and giving away lovely shopping bags sewn from repurposed fabrics, or stuffing them into people's letterboxes in a guerrilla delivery action. The community is able to contribute, interact and converse via a web-based forum and their website[35] offers all sorts of interesting and useful stuff, from a simple bag pattern to a draft letter that might be submitted to your local charity shop in the quest for more bag materials. There is even a label design for folks to download and print if they so desire.

This bagging community is not restricted to participants in the United Kingdom, but is a phenomenon that is spreading worldwide. The bags are simple to sew, are pleasant as well as useful projects to work on while enjoying the companionship of the sewing circle, and unlike quilts or other precious handiwork tasks they won't be spoiled by the occasional splash of tea or wine.

  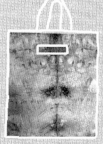 

Used jumpers (sweaters) can be made into sofa cushions; cut off the arms, sew across the remaining holes and the neck and add a decorative row of buttons to the waist hem. Use the sleeves to make lovely fingerless gloves, 'armies' or legwarmers. A jumper can also be sewn into a shopping bag with nice sturdy handles. Give it a hot wash when things like jeans are being laundered and, providing the jumper was not knitted from superwash wool, the bag should become nicely felted by the brutal wash treatment. Stuff it with newspaper while still damp and form it into the shape you would like it to be, remembering that wool has a memory for the shape it is dried in.

left: a dress constructed from a mixture of Milkymerino™ and vintage kimono silks (model: Violette Flint); right: detail

Some simple tricks to freshen tired garments

- Add appliqué, patches, lace or extra pockets.
- Cut off or slightly shorten the sleeves.
- Over-dye – soak an old stained white T-shirt in cow's milk overnight. Dry the shirt in a shady place and allow it to cure for a couple of days. Then rinse it lightly in rainwater and dye it using any of the techniques discussed in Chapter 10. The beauty of plant-dyed fabrics is that they can be refreshed in the dye-pot over and over again, unlike synthetically dyed cloth that often shows a resistance to further dyeing because the threads have been so firmly coated in the first instance.
- Shorten a skirt, then use the piece you have removed to lengthen another garment. Consider adding interesting detail by sewing masses of buttons over a skirt. Op-shops and thrift stores always have a wonderful selection of buttons to rummage through and the chase for something special is half the fun. Alternatively buy pre-loved strings of beads and take them apart, mix them up and stitch them around the hem of the garment in a cheerful blur of colour. Layered combinations of beads and buttons can produce marvellously textured effects.

◀ Violette Flint (Australia) wearing her hand-sewn patchwork dress

Reconstruction ideas

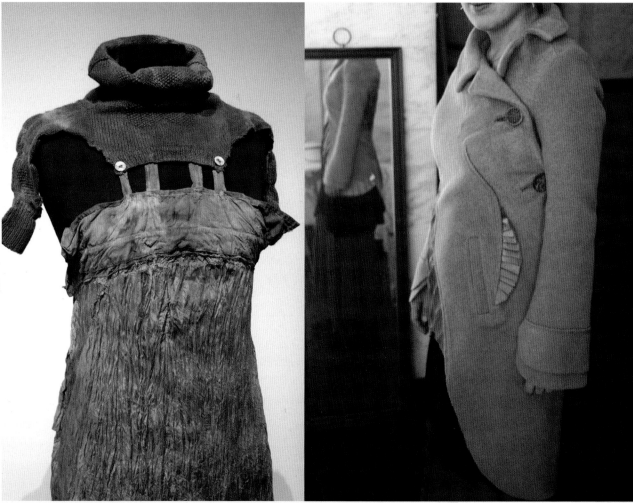

a baby's jumper was turned upside down to form the collar for this dress

Cory Gunter-Brown (USA) reconstructed wool coat

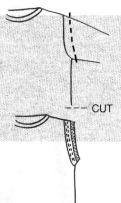

- **A sleeveless top** can be made from a short- or long-sleeved T-shirt, retaining the collar but cutting away the sleeves at an angle. Make a feature of the shoulders by cutting along the sleeve seam to the underarm, then folding the sleeves across the top of the garment and stitching them down using a *boro* or *kantha* technique.

- Make a **cross-over wrap top** from a long-sleeved T-shirt by cutting across diagonally on the front of the garment, then cutting the sleeves to half length and extending the fronts using the leftover pieces from the sleeves.

- **Turn a tablecloth into a skirt** by cutting a hole in the middle (determine the size of the hole by measuring around your waist), stitching a triangular gusset into a slash that extends from the hole and adding a button or two for easy doing up. Alternatively make the hole just big enough to step into, add a tube around its edge and insert a piece of elastic or a drawstring to keep the skirt in place on your body. I like to make the waist hole slightly off-centre on the tablecloth as it makes for a more interesting skirt that can be layered over other items.

- **Shift dresses** found at op-shops can be used for all sorts of things. Cutting up the back of the garment and adding a few pockets can make it into a handy pinafore. If the shift is a size too small, unpick the side seams and insert strips of fabric to expand the dimensions. A shift dress of the right size but of a horrible fabric or colour that you wouldn't ordinarily wear can be used as a basic pattern from which many variations could be drafted.

- A knitted jumper that no longer entrances in its present form can be turned into **a very comfortable skirt**. Begin by cutting the sleeves short (to about T-shirt length) and put the cut-off pieces aside for making 'armies' or fingerless gloves. Next cut along the top of the jumper, from the middle of the side of the neck straight down the top of the arm as shown in the diagram. This opens out the jumper and becomes the hem of your new skirt. Leave it short or enhance the hem by the addition of a ruffle (stretching out the knit as you stitch the ruffle on) or add a straight piece from another jumper. The waist of the original jumper is the waist of the new skirt; it's just that the jumper now hangs down from the waist rather than being worn on the upper body. Depending how firmly the waist fits, elastic or some other closure might need to be added.

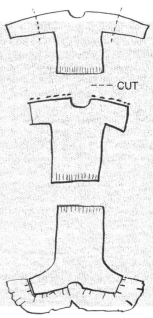

- **Layer two similar T-shirts** of contrasting colours one over the other, stitching them together carefully so that some of the colour from the inner shirt is visible from the outside. Endless variations can be played with by cutting holes and patterns into one of the shirts and then overlaying the uncut shirt with the pierced one.

- Alternatively, **combine two T-shirts** to make one wrap top, either by using two different colours (complementary or harmonious) or using the same colour throughout.

- Two jumpers of similar fabric weights can be combined with an old silk shirt to make a **large comfortable cardigan or knitted coat**. Cut both jumpers up the middle of the front only. Trim the sleeves from one and open it out in the same way as for the skirt described on page 199. Stitch the jumpers together at their waists. Cut the sleeve of the top part of the cardigan off at the elbow then take the sleeves that were removed from the first jumper and stitch them onto the shortened ones, making a two-colour bell sleeve. Use the chopped pieces of sleeve to make pockets on the cardigan. Take the silk shirt and cut off the sleeves, keeping them to be used later as cardigan pockets as well. Next cut the shirt into strips about 10 centimetres (4 inches) wide and hem the strips down one side. Make sure you leave enough allowance when you cut for a seam to join the silk to the cardigan. The collar of the shirt can either be aligned in the normal collar position or for added interest placed on the lower hem of the cardigan. Gather the silk slightly as you stitch it onto the knit cloth so it forms a ruffle. Stitch two pieces of silk at the front for a bow tie, or add a collection of old buttons to one side and a strip of buttonholes from an old shirt down the other.

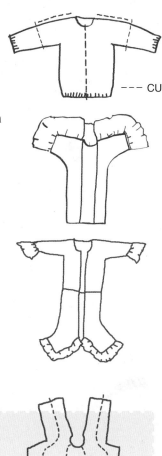

--- CUT

Tip: If you cut into the shirt from the middle of the lower back hem you should be able to get a good long strip that wanders along the lower hem to both sides and then up to the collar in the middle.

- An amusing way to **repurpose an old jumper** is simply to take a pair of scissors to it and cut slashes all over it about 7 centimetres (3 inches) long and 2 centimetres (1 inch) apart. A really brutal hand-wash in hot water with some soap will help to firm up all the edges of the cuts and you will find the garment will stretch quite a bit. Wear the lacy jumper over a long-sleeved T-shirt in a contrasting colour so that you can see the effects of your work. (This makes a much looser garment than the version described on page 162.)

🌸 **Old jeans**, being made of durable denim, can be repurposed in many ways. Unpicking the leg seams and realigning the pieces can produce a wonderfully comfortable skirt without the tedious insertion of zips or buttons. Insert gores made from contrasting fabric pieces (or another pair of jeans) to add width to the lower edge of the skirt. Hand embroidery that is allowed to wander randomly across the surface will serve to unite the different fabrics as well as personalising your design. Choose a larger pair of jeans to work with in the event you wish to make a dress and add a pair of straps crossed over the shoulders to keep it from falling off.

Violette Flint (Australia) dress from an old pair of jeans, hand stitched with 1930s vintage linen thread (detail)

Violette Flint (Australia) skirt patchworked from jeans

🌸 Give an old garment a completely new look simply by **changing the buttons**. Make sure the new buttons fit the buttonholes before making the leap. Stitch on a range of different coloured buttons for fun.

🌸 Whatever you do, take your time. Relish the meditative process of stitching. Working on your clothes should be an enjoyable and satisfying experience as much as a means of saving money or simply being different from the rest of the pack.

Making a pinafore from an old shirt

Taking a generously sized shirt (cotton, hemp, linen, nettle or silk), remove the collar and sleeves. Set them aside for later.

Then cut along the placket (that's the bit that holds the buttons and buttonholes at each side), keeping fairly close to the seam. If the placket facing hasn't been stitched down you might like to do this before you cut.

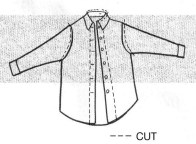
--- CUT

When you reach the top of the placket, cut along the shoulders until you reach the top of the sleeve.

Open out the shirt and spread out the bits that were once the front and the sleeve tops. The back of the shirt has become the front of your new pinafore.

Now take the sleeves that you set aside before. Choose a cutting point, avoiding the placket and button, and open out each sleeve. Stitch the opened-out sleeves to the front of the pinafore (cuffs pointing upwards as shown) for a pair of elegant patch pockets.

Cross over the straps that were the shirt placket and button them at the point where they overlap so that you can easily get your head through the remaining space. Stitch each placket end to the top of what was once the opposite shirt front (see diagram).

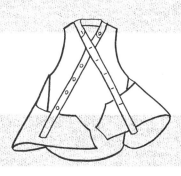

Stitch the tops of the two former shirt-fronts together (or add a button and loop closing).

You can choose to hem your pinafore or revel in the lovely frayed edges. For variations, add patches to the surface and as many pockets as you like.

I like to add silk fragments and then bundle the whole garment into a dye-bath for a complete transformation.

Tip: Leftover collars can be dyed and beaded and used as an amusing accessory. Silk cuffs cut from shirts can be similarly transformed, keeping their buttons in the original position to fasten the item in place on the body.

More ideas

Tear (or cut carefully) the button plackets from the front of otherwise unuseable shirts, making sure there is a good section of fabric left as your seam allowance. Stitch these pieces onto a backing cloth, overlapping as you go, making a layered cloth ornamented with many buttons. This technique is also lovely using just the buttonhole plackets. Either way it does involve some dedicated collecting!

Crochet many small shapes in various colours. Stitch them all together to make a garment or a scarf. Be brave and play with organic shapes, just joining them together where they meet. There may be spaces in between, in which case you will have made lace.

Braid strips of rag together and stitch them into a garment, randomly following your body shape (or using a dress-form or dummy).

Cut the soles from a worn-out pair of tennis shoes, blanket stitch around the edges and then crochet a shape to fit your foot – slippers, boots or sandals.

Make a skirt, a shift or even a curtain by joining a collection of cotton lace doilies together.

Make string and knit or crochet a garment, following your intuition. If you run out of string, just twine some more on to the raw end.

Sew up an old singlet (vest) along the bottom hem. The shoulder straps make perfect handles and knit singlets make bags that are wonderfully expandable.

Collect all the shreds from your overlocker and the off-cuts and thread trims from other sewing. The longer bits can be used for string-making and the short or fluffy pieces tossed onto the surface when building up a felt for interesting texture and pattern.

nine lives shooz by Violette Flint (Australia)

A garment by Trish Edwards (Australia) being constructed on a dress-form during a costume workshop at the Geelong Textile Fibre Forum 2008. The skirt of the dress is formed from several shirts, the bodice from a cotton-knit top to which the sleeves of another garment have been added

Join three or four shirtsleeves (opened out by cutting up the seam) to form an amusing skirt. Leave the cuff plackets and their buttons in place as a means of getting in and out of your skirt.

Old jumpers can be stitched together in layers or felted using a hot wash in a machine and cut into strips. Knit them together on pieces of dowelling to form a mat. Fling the mat into another hot wash (perhaps with something sturdy such as jeans) so that it felts together even more and becomes a good deal thicker. Before beginning the process it is worth checking whether the jumper is made from 'superwash' wool; if this is the case it simply won't felt.

Liz la Sorsa (New Zealand–USA) work in progress – a child's jacket being constructed from an old soft nappy, dyed with plants and trimmed with soft wool felt

Leftover sleeves can make very handy fingerless gloves for those cold winter mornings in the garden. Loose or sloppy sleeves will need to be taken in a little so that they fit firmly and don't slip down the arms. Simply cut a thumb hole about 5 centimetres (2 inches) from the edge of the cuff. Hem the edge using an overlocker or do it by hand with blanket stitch. If the thumb hole has accidentally been cut too large, make a feature of it by crocheting decoratively around the edge. Decorate the glove as desired with other fabrics or with beads or buttons – making sure they don't get in the way of any practical uses.

The ragbag

Too often the ragbag becomes a kind of pit of despair where old clothes and newly single socks go to die amid bundles of scraps from the making of clothing. To stop this from happening I try (not always successfully) to group my fabrics in a number of smaller receptacles according to colour, age and durability. If they are put into small stacks with their friends and tied up with a nice piece of string or ribbon it makes everything much easier to find as well as being decorative. One could even make an artwork of the rag collection by storing it in a glass-fronted cabinet with lots of little pigeon-holes or small compartments. The truly ragged stuff that doesn't have much of a future can be relegated to the place where cleaning supplies are kept.

And who says you can't wear odd socks together?

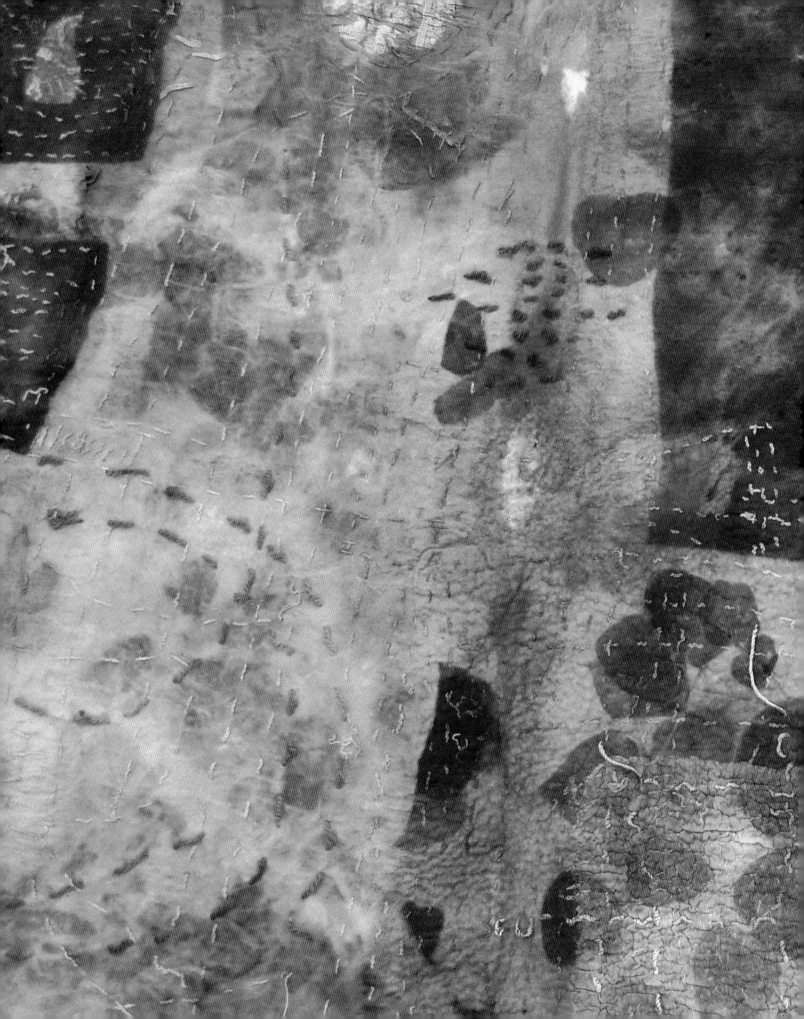

chapter 9

When all else fails
patching, piecing, felting and twining

A change is as good as a feast.

Gerald Durrell[36]

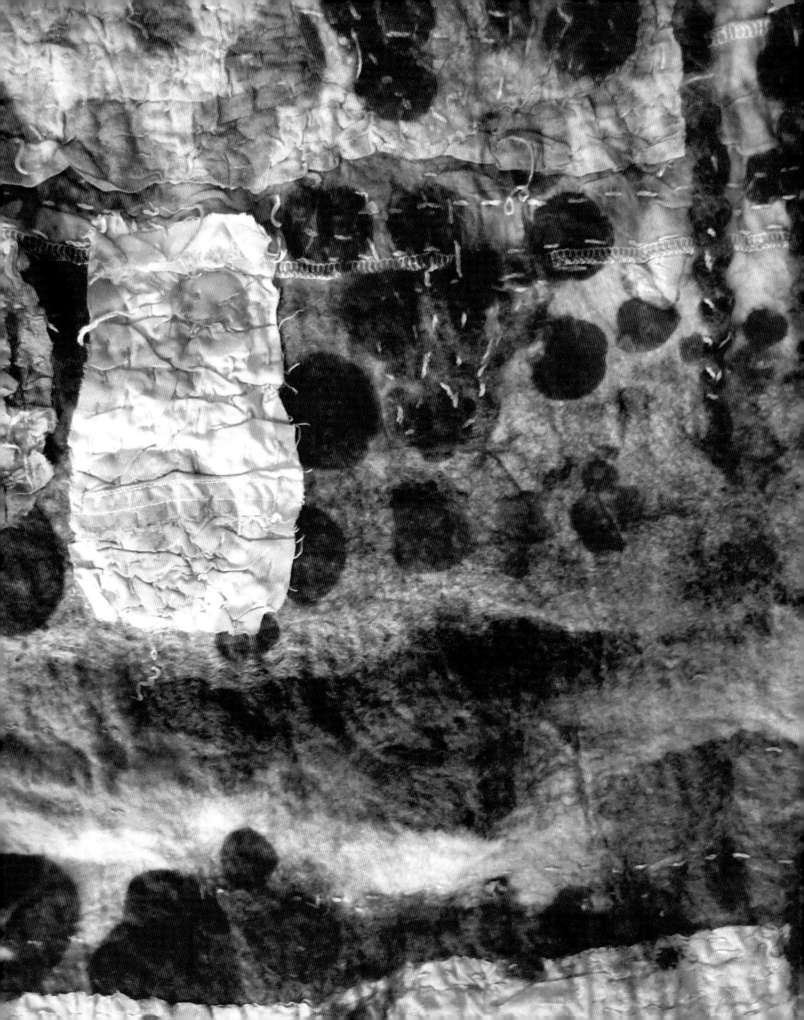

When your garment or other domestic textile has become so worn that it can no longer be repaired to sustain its original purpose, it can still be useful. The best pieces can be used for patchwork and appliqué while the remainder of the item can be cut into long strips for rag weaving or shorter strips for making string.

Extremely fragile pieces of cloth can be used in the making of felted quilts as the wool used for batting helps to bond any soft frayed bits firmly together.

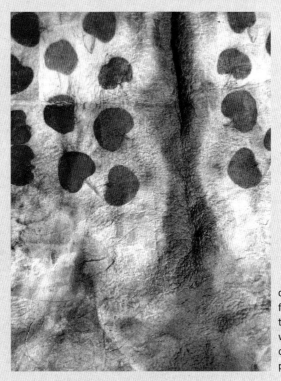

detail of felt fabric formed from layers of wool sliver together with odd-sized woven fragments that have defied inclusion in traditional patchwork

If you really cannot envisage a use for an old piece and it is so soft that it is beyond redemption even as string or a cleaning cloth then it can be shredded to use as cushion stuffing or even used to make paper. By making the best possible use of our resources in this way we can play an active part in reducing landfill while at the same time finding pleasant occupations, making our homes more comfortable and as a bonus, saving a lot of money. I find it so much more satisfying to make things for myself than to spend hours trawling shops in search of the perfect item.

◀ silk scraps can be incorporated into felt – stitch them lightly onto the piece at 'pre-felt' stage before fulling to completion (see page 213)

Invoking the magic of wool

Wool is one of the most exciting of the natural fibres as it offers so many possibilities. It can be spun, knitted, woven, crocheted and best of all felted. As discussed earlier, the individual filaments have a structure that resembles fish scales and when the fibre is warmed and wetted these scales swell and stand away from the cortex. This allows dye particles to bond firmly, as when the wool is dried the little scales flatten again and lock down over the colour. It's a simplistic explanation, but wool really does need to be heated to pick up the strongest and most enduring colour. The exception is when wool fibres have been so processed by stretching and by application of chemicals that the original attributes have been lost. The destruction of fire-resistant properties by the application of the chemicals used in the 'superwash' process is but one example. The scaly nature of wool is also vital to the process of making felt, as it allows the wool fibres to slide closer together when friction is applied and also means that as the wool fibre dries the scales lock down over whatever other filament is in contact with them, gripping it firmly. This is why woollen objects remember the shape in which they are dried, why jumpers pegged on a clothesline have marks in them until the next time they are washed and also why we can use wool sliver to bond pieces of soft fabric together. Considering the relative rapidity with which a large piece of felt can be constructed when compared to a textile of similar size laboriously woven by hand, it really is a most efficient method of constructing cloth. Felt is also extremely versatile as it can be made flat (two dimensional) or created as a three-dimensional object without visible seams if so desired.

scraps of loom-waste bonded onto wool felt

Anyone who has inadvertently washed their favourite non-superwash-treated merino jumper in the hot cycle of a washing machine will have a very good idea of the conditions required to make felt. They will also have discovered that felting is an ongoing and irreversible process and that (contrary to popular belief) no amount of soaking and stretching will ever restore the garment to its original condition.

When moisture, heat and friction are applied to wool it will inevitably felt. Miraculously it only felts on the sheep's back when the animal is ill or under some form of stress – even though unwashed lanolin-rich freshly shorn wool can be quite easily felted with a little water and hand rubbing.

We can use the bonding qualities of wool to embed or affix other materials as well. Lightweight fabrics can simply be felted directly into the surface, resulting in a laminated textile. Heavier fabrics benefit by being stitched into the surface of a pre-felt,[37] as a small quantity of wool fibre will be pulled through each needle hole along with the thread each time the fabric is pierced to form a stitch. Such fabrics will bond to the felt only at the point where the stitch thread passes through, creating a rich and interesting surface texture. There is a simple way to tell if your fabric needs to be stitched to the surface. If you can see spaces between the weave when you hold the cloth up to the light then it is not strictly necessary to stitch before felting, although any fabric can benefit from being sewn into place. Attempting to breathe through the fabric is another means of assessing the density; however, if it is an older or dusty piece I'd avoid doing that in case of breathing in dust, spores or micro-fauna. If the fabric blocks the passage of light it probably needs to be stitched down prior to fulling the felt.

'Fulling' is the process of repeated rolling and stretching that takes a felt from the semi-felted or 'pre-felt' stage to a firm, resilient fabric. If wool fibres can be pulled from the surface then the fabric is not considered to be fully felted or fulled.

Making a simple piece of felt using wool sliver

When wool is processed by a mill it is scoured, gilled, combed and carded prior to spinning. Scouring means that the grease and dust are washed from the wool. The processes of gilling, combing and carding involve the removal of any vegetable matter as well as straightening and aligning the wool fibres in what is variously known as sliver or roving. At this stage the wool is perfect for felt making.

To make a simple piece of felt, all you need to do is lay out wool sliver evenly across a piece of old sheeting, bamboo blind or bubble wrap, pulling it gently from the end of the rope of sliver while holding the bulk of it well back with your other hand. Wool fibres are stronger than fibres of steel of the same diameter and are quite difficult to break, but if you simply pull from the end of the sliver with finger and thumb then wisps equivalent to the length of the staple of the fleece (usually 8–10 centimetres/3–4 inches) can be plucked without difficulty.

pulling off strands of sliver

Having put down one layer of wool, add another layer at right angles to the first. Wet the wool by sprinkling it with lukewarm water in which a bar of soap has been lightly swished a couple of times. Soap (or mild detergent) is necessary to break the surface tension of the water and help wet the wool, but bubbles should be avoided as they inhibit the felting process.

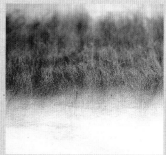
arranging layers of sliver at right angles to each other

There are various ways of applying the water. Some felters like to use a small watering can or spray bottle, others dribble water on with a sponge. (Had she been a feltmaker I suspect my grandmother would simply have taken a mouthful of water and sprayed the wool in the same way she expertly misted her laundry while ironing. Be warned, your children will be horrified if they catch you trying this.) A piece of netting laid over the wool can be useful too; rub the wool gently (through the netting) with a scrunched recycled plastic bag, or pat it with your fingers to help the water work its way through the fibres.

applying warm soapy water

When you are satisfied the wool is wet (dry parts will look quite white in comparison to the rest, assuming you are not using coloured wool), gently roll the wool up using your support (cloth/bamboo blind/bubble wrap). Roll the bundle gently back and forth a few times, then carefully open it up and roll it from another direction. Continue to roll with regular inspections until the felt feels firm.

The edges of the felt can be worked with a little soap to help them become sturdier. These sorts of felts can be made on the kitchen sink or in hotel bathrooms while travelling if the mood strikes. If you plan to join many small felts together, stop felting at the pre-felt stage, when the wool will hold together if picked up but is still soft enough that fibres can be picked from the surface.

On the other hand, fully felted small pieces can be joined by blanket stitching around the edges and then joining the pieces with crochet.

Tube-felted scarf

Here is a method of using either commercially needle-felted wool pre-felt or your own pre-felt to use up the scraps from the sewing room floor that are too small for patchwork, appliqué or string making. You'll need a piece of cardboard no more than 90 centimetres (1 yard) wide or high (that is, no further than your arms can reach for ease of stitching) and a piece of wool pre-felt twice the size of the piece of cardboard plus a little for the overlap. The cardboard forms a resist so that the actual piece of felt resembles a flattened tube or cylinder. The ideal cardboard size for ease of handling is about 30 x 60 centimetres (12 x 24 inches).

Wool pre-felt can be purchased or made by laying out wool sliver and felting it until it is strong enough to hang together but not so tough that fibres cannot be pulled from the surface.

even brightly coloured scraps can sing happily together when unified by felt

You'll also need a piece of silk or cotton muslin of the same size. If you are using the cardboard size recommended on page 215, then the cloth needs to be 70 x 60 centimetres (28 x 24 inches) in size to allow for the overlap; it could even be an old scarf. If the scarf is fairly sturdy you may like to cut a few holes in it to allow the other fabrics you will be using to peep through. You might also like to use some wool sliver for added effect. A needle, scissors and a selection of threads will complete your materials kit.

Method

First spread the piece of muslin on a table or other large flat surface. Then lay scraps of fabric (wrong side up) on the muslin. You can also lay strings of old beads or sprinkle a few old buttons on the muslin if you wish. Spread a few wisps of wool sliver over the fabric ends, beads and buttons, then lay the pre-felt over the top.

Now place the sheet of cardboard in the middle of the fabric sandwich and carefully fold the fibres around the cardboard as shown. Tack the layers together where they overlap. Continue to stitch and embroider through all the layers of cloth and wool but not the cardboard – its role is to act as a resist and to keep the sides of the tube from being joined together inadvertently – until you have a richly textured surface. Take your time and enjoy the stitching journey. The beauty of this method is that the more uneven the stitches are, the more interesting the finished surface will be. Quite sturdy fabrics that would not ordinarily be used in making laminated felt can be bonded using this technique, simply because each time the needle pulls thread through all the layers it is also pulling the fine wool filaments, which when felted will give your reconstructed textile a wonderfully textured but even (due to the placement of the stitches) crumpled look.

profile of layers
(from bottom): muslin, scraps, sliver, pre-felt

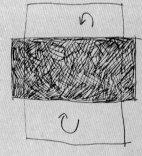

Try using a variety of materials for the stitching, such as embroidery threads, fine strips of torn cloth (including the selvedges), wool yarns and various combinations of filaments.

When you are satisfied with the texture and are sure the small pieces are all fixed, prepare some warm water with a little wetting agent (soap or an eco-friendly detergent). Place your work on a piece of plastic bubble wrap and then sprinkle it with the detergent and water mixture. Gently rub the surface with a piece of recycled plastic, patting down any loose ends with your fingers. You will be able to feel the various fibres beginning to bond together. When you can no longer easily pull pieces away it is time to cut the tube (for the shape you have made around the cardboard is a tube of sorts) away from the resist.

Cut in a spiral, using the pasting lines on the inner cardboard of a toilet roll for guidance if need be. In this way you will end up with quite a long and beautifully textured scarf when the felting has been completed.

To finish the felting process, either toss the piece about, warming it in fresh warm water from time to time, or keep rolling for a crisper finish. Felting can be hastened by kneading the scarf gently as if it were dough. Occasionally smooth the work using the piece of recycled plastic that you used to apply the water. Another useful tool for smoothing is a small block of wood that has been enveloped in recycled bubble wrap secured with a rubber band or two.

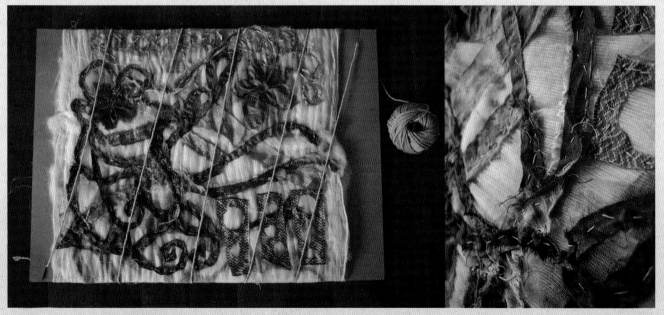

marking the cutting lines 　　　　　　　　　　　extra bits randomly stitched onto the surface

When the surface begins to develop a slightly wavy and 'bubbly' texture, the felt has been sufficiently fulled and the process is complete. Rinse the scarf, spin out excess moisture and hang in the shade to dry. If the soap or wetting agent you have used has made the wool feel harsh, soak the scarf in a little water to which a splash of vinegar has been added. This neutralises the alkalinity, having much the same effect as conditioner has on hair, and is far kinder to the cloth than so-called fabric conditioners.

Once dry, further embroidery may be added as well as more beads and buttons if desired.

If you have made the scarf in neutral or undyed colours then it can be enhanced by over-dyeing using colour derived from plant dyes. Alternatively make the scarf using pre-coloured scraps.

Some felting tools from the kitchen

Old onion bags made from plastic netting are handy when scrunched up and used to firm the surface of felt.

The big bags that oranges are often supplied in are very useful for wetting out felt. Lay the orange bag flat across the work and then dribble water over the top, using either a bowl or perhaps an old milk bottle that has been pierced a number of times through the lid. Then gently begin to rub the surface of the felt through the orange bag, using a piece of scrunched plastic packaging or a netting onion bag that has been stuffed with other netting onion bags. The friction provided through jiggling the various pieces of netting together encourages the felting process to happen more quickly.

The ridges on the kitchen sink are very useful for rubbing the felt against to make it nice and firm. The felting process is also quite useful for cleaning the kitchen sink, so it is a win–win situation.

Felting around cardboard is fun, even though the resist does eventually become a soggy mess (use it in the garden as mulch). All sorts of interesting shapes can be cut from card and used to make cushions, bags and wild hats, depending on how the object is cut (or not cut). Cardboard has the advantage over bubble wrap as a resist because it's nice and stiff to begin with and then relatively easy to remove when soggy.

When it is constructed in neutral or undyed wool and fabric scraps and subsequently dyed using plants, I call this a 'landskin' scarf. The great advantage of felting around a resist in this manner is that a comparatively long felted object can be created without placing excessive strain on the back. After years of feltmaking I am finding that laying out wool (when the arms are extended in front of the body) and the physical process of wetting out large areas are both beginning to take their toll.

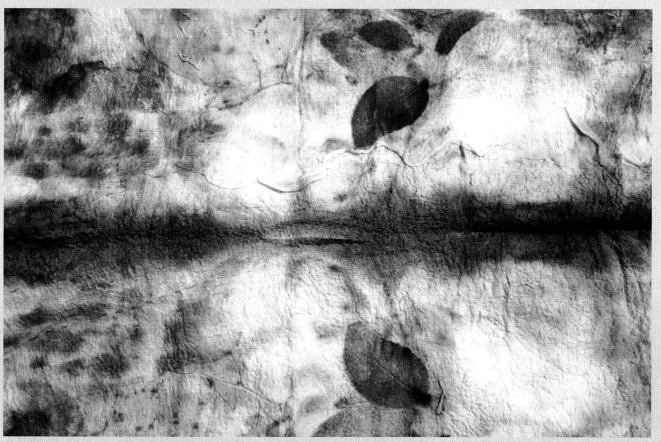

'Landskin' (detail)

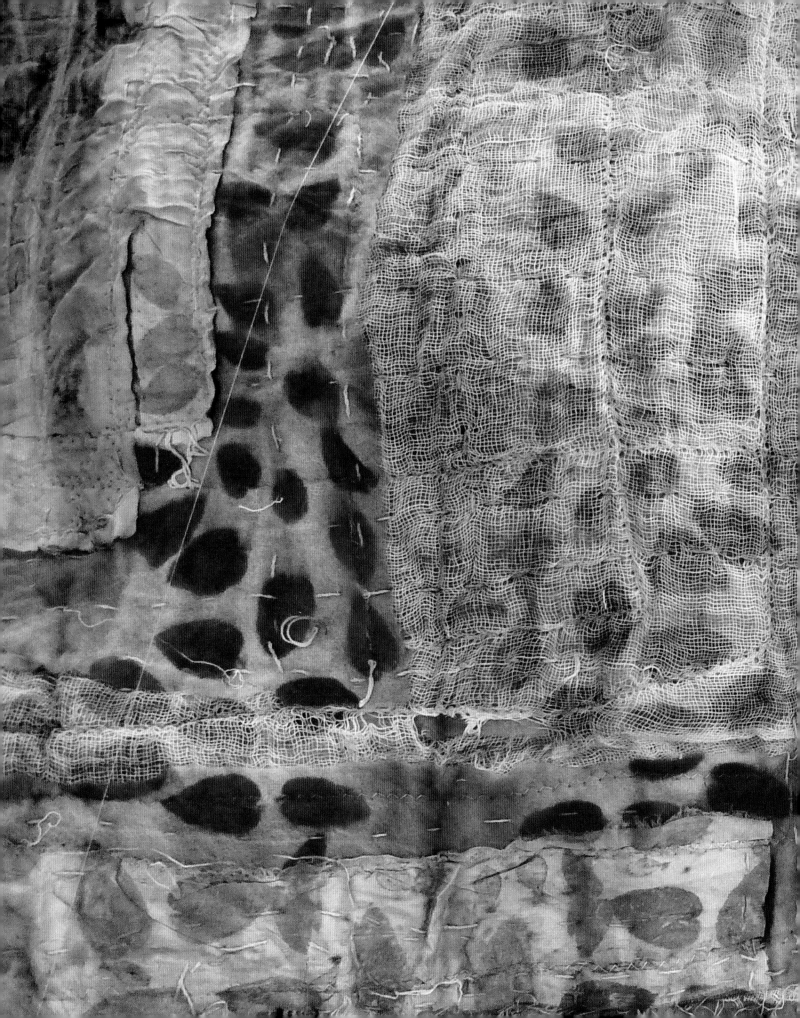

Making quilts from old clothes

I become very attached to my favourite clothes and wear them (with complete disregard for fashion) for years on end, patching and re-dyeing them as required. I sometimes think it is a pity I can't restore my body in the same way. Some old clothes hold so much meaning and memory that they become quite special. They become reminders of family and friends and places, of how one felt about life at certain times and of important things that have happened along the way, making it even harder to put them in the charity bin. Mind you, those of my garments that have been thoroughly patched probably would not be welcome there anyway. Converting such objects into hand-stitched quilts makes perfect sense.

The make-do-and-mend tradition of quiltmaking probably dates back almost as far as the invention of woven cloth, when there was a greater understanding of the effort involved in production and every fragment was considered precious.

the reverse side of an antique Uzbek quilt top, glowing like a stained-glass window with sunlight behind it

fabric pieces randomly machined and unified by the dye-bath, which could be stitched onto wool pre-felt and then lightly fulled to make a quilt ▶

a quilt may begin as a doodle

The very simplest style of creating a larger piece of fabric from smaller patches is the 'crazy' quilt, whereby random shapes are matched, overlapped, aligned and stitched together, often with some form of decorative stitching. Using an old cotton sheet as the backing on which to stitch the crazy pieces helps ensure the created cloth is reasonably flat. Without such a backing as a guide it is easy to end up with a pieced cloth that has bulges, ridges and valleys, and while this might be decorative and sculptural, it might not be terribly practical in the context of a piece for the bed or the wall.

Working with a set of repeated regular shapes can bring a pleasant sense of order to the quilt even when the fabrics are quite different. Nine-patch and simple four-patch styles are lovely examples of this. Smaller pieces are stitched together to form groups called 'blocks' and the quilt gradually pieced together. For more complex shapes the pieces would be formed around paper patterns by folding over a small hem and tacking through the paper. This had the advantage of giving the pieces a bit of structure and firmness, making it easier to stitch them together. In less abundant times when paper was not so readily available, quiltmakers even resorted to cutting up old letters to use in their work. Mostly the papers were removed before the quilts were filled and backed, however examples have been found with precious documents still stitched in.

◀ *boro* floorcloth from Japan (collection of the author)

A quilt can begin with a simple felt-pen doodle in a sketchbook, followed by a little play with the watercolour box. To take this drawing further and realise it as a quilt, the next step could be to draw it at full scale on paper and cut patterns using paper from the recycle box. Cut the fabric pieces a centimetre or so larger than the paper patterns, fold the edges over and tack them down, before joining each piece to its neighbour using tiny stitches that run along the fold of the fabric. The papers can be taken out once the whole top is pieced. Making a quilt top in this manner is a long slow process, meditative, gentle and portable, and well worth the time taken.

detail of a quilt from the Pioneer Women's Hut, Tumbarumba, New South Wales – the quilt was found at a rubbish dump

randomly joined fabric scraps for transformation with felt

Quiltmaking need not be restricted to woven fabrics. Knitted textiles produce quilts with good drape and delightful wraps can be formed from stitching together small fragments of pre-loved knit clothing.

Very old jumpers that have become pilled, moth eaten or excessively worn can be used to create the inner insulating layers of quilts, while the backs can be simply constructed from old flannelette sheets.

You can choose to sew running stitches through all three layers to bring them together or simply stitch on buttons at regular intervals. Ensure the button thread travels through all the layers and tie it off securely.

The felted quilt

One of my favourite methods of quiltmaking brings together felting, stitching and (more often than not) dyeing. I call the technique 'feltquilt' and it's a rather organic way of making a pieced layered textile. I reserve the decorative and silky fabric for the top and use pieces of old cotton sheeting for the base. This is to prevent the finished quilt from slipping off the bed, as silky fabrics on the underside make for a quilt that has a tendency to wander off to the floor at inappropriate moments.

The pieces are quilted together by sandwiching a piece of pre-felt between the decorative layer on the top and the soft sheet layer underneath. If the pieces are kept to a maximum of 30 centimetres (12 inches) in diameter they will be easy to manage and can be tucked into a bag for stitching even while one is out and about. I find it easiest to lay the work flat on a table or tray and then work running stitch through all the layers, pushing the needle against the resistance of the flat surface. I simply join one piece to the next as each one is finished, not worrying too much about how the design might look.

machine quilting two layers of cloth filled by a layer of pre-felt – the beauty of this process is that the final 'fulling' of the cloth means any mistakes or accidental rumples in the seams become melded into the quilt and simply contribute to its rich texture. I usually work by hand, which allows far greater variation in stitches and is a much more peaceful and meditative process – machine-stitched lines tend to be a little harsh

I work the piece until it is about 50 per cent bigger than the finished quilt is required to be. This includes a 33 per cent allowance for shrinkage when the piece is being rolled with warm water and a little wetting agent in order to finish felting the wool. The beauty of using pre-felt is that one needn't be too fussy about loose ends or about hemming any edges, as the wool fibres will bond with all the flapping bits during the fulling process.

The pre-felt can be purchased commercially or simply be hand formed in small sections on the kitchen sink using the warm washing-up water after the dishes have been done.

silk and wool yarns contribute to the surface of a hand-stitched feltquilt

◀ 'Red blanket wagga', a wool quilt made by the author and now in the collection of the National Wool Museum, Geelong. This quilt was built up from pieces of cloth overlaid on a salvaged wool blanket and subsequently over-dyed using eucalyptus dyes

String-making

To make string you'll need a supply of any natural fibre cloth. Old cotton, linen or silk shirts are ideal.

A small bowl of water will also be handy, as will a pair of sharp scissors.

Note that these instructions work for either hand, so instead of referring to 'left' or 'right' the references will be to 'holding' or 'twisting'.

Cut or tear the fabric into strips 5–10 millimetres (¼–½ inch) wide and no more than 30 centimetres (12 inches) long. If the strips are longer they will become tangled in the twining process.

Place a handful of fabric strips in the bowl to dampen them. Squeeze them out so they don't drip everywhere.

Take the first strip and fold it over about two thirds along so that there are two pieces of fabric hanging down from where you are holding them and the two pieces are of uneven length. This is important so that when the time comes to join more strips to the work there won't be a nasty great lump of fabric at the splice.

dress from repurposed fabrics with hand-twined string lace

Using the finger and thumb of your preferred working hand, begin to roll the fabric strip about one third along from the end, so that it is firmly twisted. Now hold that part with the finger and thumb of your other hand. Let's call this hand the 'holding' hand. The beginning of the string will be in the form of a small tight loop.

Holding the strip in one hand as described, gently roll one of the dangling bits away from you using the finger and thumb of the other hand (the 'twisting' hand).

Bring the rolled strip forward (towards you) over the other strip. The finger and thumb of the holding hand should move just slightly along to secure this twist.

Now take the other dangly bit in the twisting hand and roll it away from you.

Bring that rolled piece towards you over the top of the first rolled bit.

By repeating this action you will be creating an S twist in conjunction with a Z twist and by a miracle the string will not unravel. Keep going until the shorter end is about 3 centimetres (1 inch) long.

Take another piece of damp fabric strip and lay it over the short end so that the pieces are overlapping. Twist together as before.

You can make the cord as long as you like – after all, how long is a piece of string?

Small fragments of special fabrics can be twined into rather nice bracelets or necklaces. Make a figure of eight knot (or add a button) to the end. This can be slipped through the loop at the beginning of the string.

Lovely variations can be made through choosing different coloured fabrics (zebra-striped twine) or adding beads and seashells as you go.

hand-twined silk string

# Unravelling yarns from old jumpers for re-knitting

When I was little I watched my grandmother painstakingly unravel hand-knit woollen jumpers that were either too small or were considered for some reason to have lost their shape.

The first part of the process simply involved winding the wool into a ball, having carefully unpicked the seams and then found the yarn ends at the top of the looped fabric pieces. The winding followed a pattern that began with a simple orbit around the hand and changed by 90 degrees at regular intervals, always wound over the fingers that were holding the ball so that the yarn wouldn't be wound too tightly.

Next she would transfer the wool from the ball to a skein, winding it from her hand to the elbow on the same arm. The skein would be tied at four places using a fragment of wool in a contrasting colour and following a figure-of-eight pattern that stopped the tie from slipping.

Once the skeins were all prepared the wool would be washed. Grandmother would fill a wide bowl with quite warm water through which a bar of soap was swished a couple of times and then drop the skeins in dry, pressing them down gently. The yarn and water would be left to cool, whereupon she would empty the wash-bowl, squeeze the skeins out gently and then rinse them in cool water.

The last part of the process involved hanging the skeins over a broomstick and weighting them by placing a full food tin in the bottom end. In this way the wool was cleaned and the irritating crinkle and curl of the unravelled yarn would be removed. I go one step further and add a drop of eucalyptus oil to the rinse water to give the yarn a lovely fresh scent.

And another thing

Prevent single buttons from getting lost by keeping a needle with a long thread handy in your sewing area. The buttons can be threaded on as they are found, making a lovely string that can be worn as an ornament if desired.

Cut worn-out socks into rings, join the rings by stitching or looping and make a lacy garment – or use the rings to create an interestingly textured felt.

Open out the good bits from socks that are beyond darning and stitch them together to make a patchwork knit cloth or garment.

Sew up the bottom hem of a shirt, tie the arms together and you have a useful carry bag that can be opened to accommodate large objects, or simply stitch up the waist on an old singlet for a handy shopping bag.

Violette Flint (Australia): free-form needle lace

Join scraps of fabric with long stitches, using buttonhole stitch to create needle lace between the solid parts.

a pair of wool pilchers can be transformed into a bag. Stitch up the leg holes, add a strap of some kind, dye and embellish

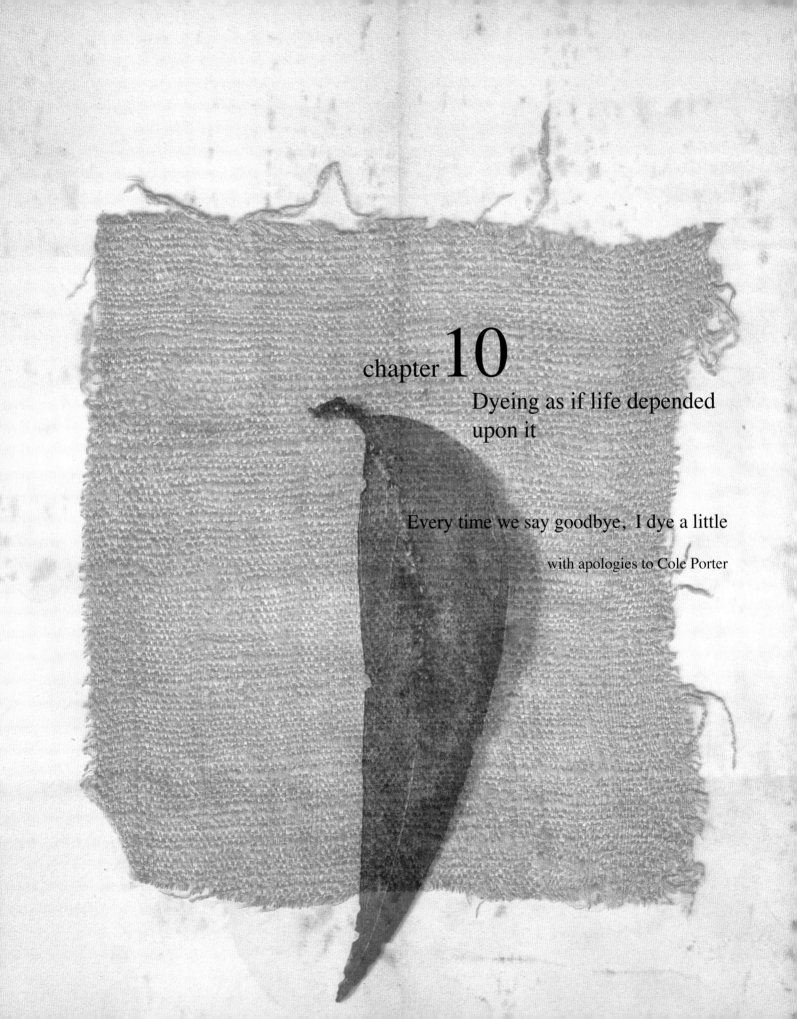

chapter 10

Dyeing as if life depended upon it

Every time we say goodbye, I dye a little

with apologies to Cole Porter

banksia flowers at various stages of development; beautiful bark on a banskia is also a source of colour

Presenting a lecture on plant dyes at a feltmakers' conference in Denmark over a decade ago I was quite viciously (but fortunately only verbally) attacked by the conference chair, who summed up my presentation by stating unequivocally that 'it was a well-known fact that boiling plants to extract dye rendered them highly toxic and thus people should stick to synthetic colours which are carefully prepared by chemists and therefore much safer' and then used her position to deny me a reply. I almost wished I had stayed at home that day! It is this sort of misinformation that has clouded the reputation of 'natural dyeing' over the years.

Certainly when the traditional adjunct mordants such as chromium, tin, copper or iron salts are used in the application of dyes, the resultant residue should in principle be disposed of at a toxic waste facility; although the unpleasant truth is that such substances have been used as mordants with gay abandon by many home-based dyers, almost certainly without proper safety procedures. *A Manual of Dyeing Receipts*, authored by James Napier in 1853 – just a few years before William Henry Perkins accidentally stumbled on a mauve dye while attempting to synthesise quinine (by unknowingly using contaminated laboratory equipment) – lists a horrifying compendium of standard chemical adjuncts that would indeed make a plant brew quite poisonous irrespective of the inherent toxicity (or not) of the species used.

How would you feel about wearing something against your skin that had been processed using compounds such as 'cyanide of potassium' or 'protoxide of uranium'? It brings to mind the legend postulating that Napoleon Bonaparte could have been deliberately poisoned using arsenic-tainted wallpaper. Whether or not the story is true, it would certainly be possible to poison a person using chemically contaminated clothing or furnishings; one could even speculate that infant mortality rates in times past might have been influenced by the way in which colour was fixed in cloth, how well that cloth was rinsed or neutralised and whether unstable toxic substances may have leached out in the presence of warmth and human bodily fluids. But I digress.

Fairly early in my chosen research path it dawned on me that in many cases where such adjuncts were used, the plant brew was acting as a kind of acidic mordant and the colour was actually being derived from the metallic salt. Consider the case of potassium bichromate, a popular (and horribly carcinogenic) mordant utilised in the making of yellow dyes. Reading the many dye recipes that recommend this substance as an assistant it becomes apparent that there must be more than a coincidental relationship between the paint colour 'chromium yellow' and those yellow dyes where chromium salts have been employed as the 'assist'.

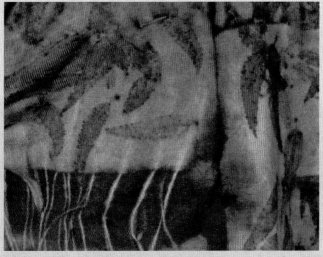

eco-print cardigan detail

Reading about the chemical make-up of that much underrated genus *Eucalyptus* (still my favourite dye source), I was astounded to discover that species of this plant contain within the leaves everything necessary to fix substantive colour on protein fibres and that applying mordants to cloth before dyeing was not necessary (unless possibly as a colour modifier). The most important adjunct in a eucalyptus dye-bath is in fact the water that is used as the substrate. Eucalyptus dyes are extremely sensitive to variations in pH as well as to dissolved salts and perform at their best in neutral to slightly acidic water. Sometimes heavily contaminated waters (such as those loaded with iron or copper salts) may, however, be used to advantage as a pre-mordant in order to influence the shade to be achieved from the dye-bath. Soaking wool or silk fabrics in such solutions (for example, bore water or seawater, or simply in some rusty-coloured water from an old tin can), then allowing them to dry before processing them in an optimally prepared eucalyptus bath offers exciting possibilities indeed. Use common sense and avoid experimenting with polluted river waters that might contain heavy metals!

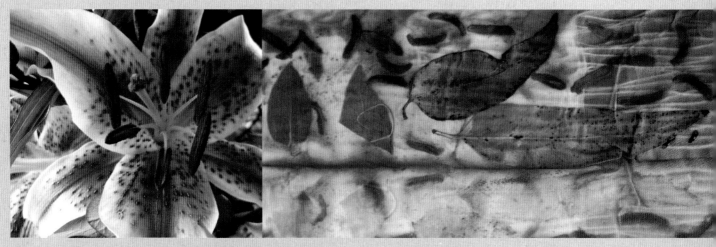

left: the anthers of lilies offer rich colour, but use them with care and bury them in the garden afterwards, as their toxicity can be strong enough to harm small creatures (such as babies and pets) if ingested; right: interesting alchemies take place when plant materials are combined. The golden marks at bottom are from lily stamens, while the green marks at top are the lily stamen remnants that have turned green in the presence of the eucalyptus leaf

The quite accidental discovery that eucalyptus leaves could be used for printing came about through the keeping of hens. They'd made their nest using old dried eucalyptus leaves, dutifully laid their eggs and then brooded quietly while I used pregnancy as an excuse to laze indoors on a few wet days (venturing out into the sodden garden only to sling wheat and kitchen scraps into their run). When I finally discovered the eggs, they had taken up the print of the leaves on the surface as a result of being slightly damp, resting on dried but still richly coloured leaves and being gently warmed by the body of the broody. All the conditions were perfect for slow and gentle dyeing! I thought a little about the way in which we dye our Easter eggs each year and mused that (given I knew that onionskins would also print on cloth) something similar might be achieved using eucalyptus. The gentle reader will be able to appreciate my utter astonishment when a fresh blue-grey leaf printed a rich deep red on a piece of silk and an even more intense red on a swatch of wool. Further thought led me to the conclusion that as this print occurred in the middle of a cloth bundle to which only steam had penetrated, the resulting colour was likely to be the outcome if the eucalyptus leaves were to be processed in pure water. I was right, but only partially, and must now admit that the theories published in my Masters thesis were somewhat incomplete – although on the right track.

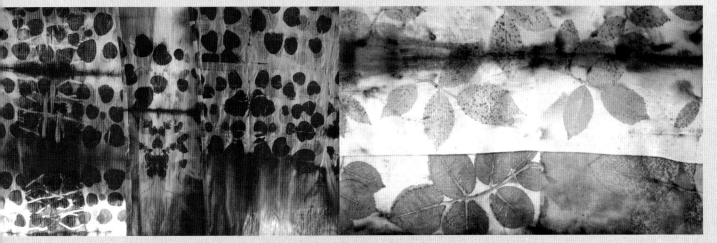

left: *Eucalyptus cinerea* ecoprint on wool; right: rose leaf on Milkymerino™ jersey

The eucalyptus is indeed an astonishing genus and the more I work with it, the more I find there is to learn. I now know that while the first eco-prints from a eucalyptus leaf will give a very good idea of the colour to be had from the first boiling, the second print can be quite a different colour, such as lime green (after an initial orange or red). I have also found that when the leaves have been boiled for 45 minutes or so (to extract the brightest dye) they can still be used for printing, again with often surprising results.

Eucalypts also behave differently when processing temperatures are varied – the colour extracted not only depends on the heat available at the time of extraction but also on the heat (or lack of) during application of the dye to the cloth. For example if a eucalyptus dye-bath were allowed to cool overnight before immersing the cloth to be dyed, the resultant colour would not be as bright as if the cloth had been dropped into the hot, fresh dye. These possibilities – combined with the fact that every part of a eucalyptus gives some sort of colour, that fallen dry leaves yield very different results to fresh picked (as well as sometimes dramatic variations in dyes from leaves depending on which part of the tree they were sourced from) and also that the intensity of each season of the year makes a difference – would indicate that a dyer could quite contentedly dedicate the rest of his or her life to simply working with one tree. The changes would vary from the subtle to the dramatic as the tree grows from infancy to maturity. There would be bark shed at different times throughout the year, sticks and twigs for the tying of bundles, leaves at all stages of desiccation on the ground below and not only a range of fresh leaf colours but also a range of leaf shapes as well, as eucalypts have a curious habit of producing juvenile leaves even when quite old, in response to pruning or other trauma. The buds, bud caps, flowers, seed caps and gumnuts all contain colour. Even the sawdust from the wood can be used successfully in a dye-bath.

While the eucalypts are rather curious in that the dyes they produce contrast so dramatically with the fresh foliage, they are not the only plants that will print onto cloth. Most plant species will make a mark on protein fibres provided they are bundled firmly enough and given time and the appropriate conditions for the transfer to take place. The wonderful thing is that those noxious mordants so liberally used in the past are not really necessary at all when silks or wools are being dyed, and provided that plant fibres are coated in some form of protein or used in conjunction with scrap metal, they too will accept colour well without other interventions.

Although some of the information presented in this chapter was explored far more extensively in the book *Eco Colour* published some years ago, I felt it important to include here a section on plant dyeing written specifically for the benefit of readers wishing to undertake adventures in dyeing and

Eucalyptus cinerea eco-print on wool

re-dyeing their clothes (as opposed to people wishing to reproduce precise colour samples for specific purposes). I've also found that while the big discoveries in the world may have been made and one can no longer sally forth like Dr Dolittle and aspire to discovering new wonders such as floating islands, there is indeed an abundance of smaller treasures still left to uncover; certainly I have learned much since writing the earlier book.

Bundle dyeing

The simplicity and beauty of the bundle-dyeing method (which I had really only begun to touch upon then) will be explored more fully in these pages. Although I am familiar with a wide range of dye applications I am finding the magic of hot-bundle dyeing so seductive and exciting that it has become quite my favourite way of applying colour to cloth. It uses relatively little plant material and almost always results in some delightful surprise in addition to familiar and anticipated marks. It is also a relatively safe means of dyeing garments, despite the evidence that cooking cloth can often lead to disaster. For example, even though a hot dye-bath presents all the conditions necessary for felting, the fact that we are immobilising the object by tying it firmly will prevent the movement-induced shrinkage we could otherwise expect in a woollen garment.

Bundle dyeing also requires smaller pots, less water and less heat energy than would be used for evenly dyeing lengths of cloth after processing and

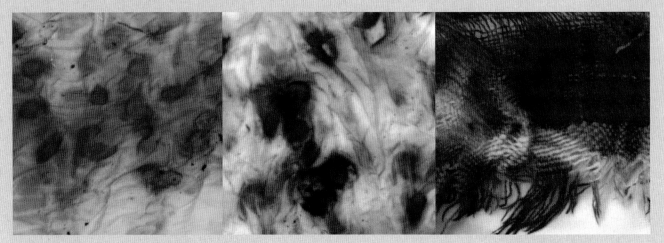

these images illustrate the possibilities for beautiful patterning when madder (*Rubia tinctorum*) is used in a non-traditional way, bundling and contact dyeing rather than creating a 'soup' extracting a dye. It is by far the most efficient, safe and effective means of colouring fabric.

Dealing with faded glory

Garments we love and wear frequently will gradually but inevitably fade as the colours on the surface become bleached by sunlight, are abraded by rubbing (even against other pieces of clothing) or change by being inadvertently attacked by the chemicals used in the cleaning process. 'New' clothes that have been constructed from a selection of pre-used items might sometimes benefit from a bit of extra colour too. An over-dye can be used to unite dissimilar fabrics, to cover stains,[39] in some cases enhance them and (if you're using eucalyptus) to quite literally freshen up the fabric and give it a non-toxic anti-bacterial treatment. Patches and darns on well-loved clothes can become less obtrusive with a dip in the dye-pot – although these days darning can be a decorative feature as much as a practical maintenance skill.

paeony petals used to dye silk

You may simply wish to apply colour to garments made from new fabrics, or to newly purchased clothing in order to give some individuality to your wardrobe.

There are many ways you can go about this. Dye colour can be applied hot or cold or simply by using a warm steeping process. You can bundle your garment around leaves and bark and dye it that way by heating it in a pot, burying it in the compost heap or stuffing it into a jar and leaving in the sunshine. Or you can simply take a small hammer and beat colour in directly from leaves and flowers. You can make a dye solution and process the whole

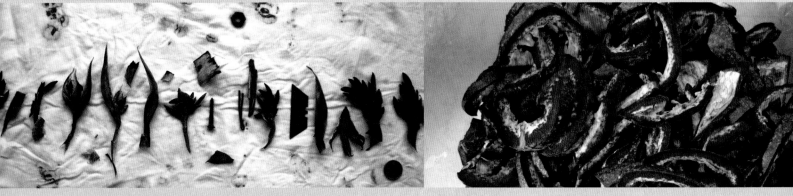

organic T-shirt showing rust stains and windfall flora ready for dyeing   passionfruit skins yield rich colours

garment so as to have a flat and uniform colour application. Whichever method you choose, the one guarantee is that your 'new' garment will be different from anything that anybody else is wearing. This is the magic of plant dyes; they reflect the personality of the dyer and the bio-region where the plants are gathered. When applied to a revitalised garment they add to the layers of narrative already embedded in the surface and continue the development of the story. Many dyes (such as eucalyptus and indigo) will even strengthen the fibre as they coat it with a firmly bonded layer of colour.

Essentially, the most important thing to understand about the use of plant-derived dye colour is that textiles for clothing more or less fall into three categories: protein, cellulose and synthetic. Each category comprises a broad spectrum of different fibres but the three distinctions are all you really need to know when dyeing.

Most plant dyes are naturally acidic in nature, with the exception of lichen dyes (slow-brewed using the ammonia method) and also indigo (whether naturally brewed or synthetically coaxed into life). However, the quality and provenance of your reticulated water supply will always have an influence on the dye-bath and, depending on the pH of the liquid as well as the invisible dissolved substances hidden in it, can dramatically affect the dye outcome. If the water is neutral or slightly acid, you'll usually get the best and brightest colours. On the other hand, water sourced from bores can often give very exciting and unexpected results. On a research trip to the Flinders Ranges (in the north of South Australia) I was delighted by the copper-rich bore water available at Wirrealpa Station, which not only left a permanent mark on bathroom fittings but also enhanced greens and golds in the dye-bath and seemed to have particular affinity for cellulose substances. The water also appeared to coax extra depth from the found iron fragments that were used to darken some of the dyes.

Building bridges – is a mordant needed?

Generally speaking, protein fibres have a natural affinity for plant dyes. Put very simply this is because the molecules that make up proteins (and alkalis) have a positive bonding point available, whereas acids (in the plant dyes) have a negative. Opposites attract, so when combined these are a match made in heaven and will form what is known as a 'polar bond'.

Obviously then, although a stain can often be achieved on the surface of a cellulose-fibre textile, it won't necessarily form a permanent bond with the cloth; think about trying to push two magnet ends of like polarity together and you get the picture. So the thing when trying to dye a cellulose fibre is to provide a protein or an alkali of some sort on the surface of the article for the plant dye to bond with. The protein or alkali will bond with the garment as well as bonding with the dye, creating a kind of 'bridge'. That said, some stains do indeed offer bonding with dyes. Drips of peach juice and splashes of wine will often react impressively with applied plant dyes. Invisible stains on areas such as the underarm parts of a garment, which may have absorbed various deodorants as well as salts excreted by the bodies of wearers over years, will often take colour far more brightly than other sections. Something to keep in mind when refreshing old shirts! One way of dodging such potential 'problems' might be to cover the whole garment with deodorant; however, such chemical interventions are not desirable. In any case the underarm areas will never quite match up to the body of the garment in terms of resultant colour.

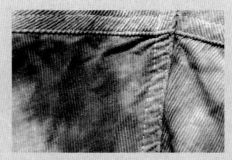

deodorant and salt from perspiration acting as a pre-mordant on this corduroy shirt

Preparing a fibre with a substance (in advance of dyeing) to encourage it to bond with the dye or to give a specific colour is a process called mordanting. A mordant can be a synthesised chemical bought from a chemist or hardware supplier (for example, alum) but it is a great deal more satisfying to make your own mordants at home. It is also rather cheaper. I usually make my own mordants using simple substances such as fruit peelings, aluminium foil and rusty nails but have from time to time used a pinch of alum per batch when preparing ice-flower dyes; however I use so little of this substance that I haven't yet reached the bottom of the 500-gram (one-pound) box I purchased in 1989. Ash from the fire works almost as well. Detailed instructions for making mordants may be found a little further into this chapter.

varying shades with home-brewed mordants

The preparation of cellulose cloth for dyeing can be a long and involved process, such as the multiple dipping and drying practised in Japan. The dyers of this nation have known for thousands of years that the use of a protein-rich mordant will intensify and help to fix colours on cloth from plant dyes. Ground soybeans were soaked in water overnight and then strained and squeezed through a fine cloth to make a milk-like solution in which the textiles were then soaked before drying in the sun. Traditionally alternate layers of ash solution and soy milk were applied, with the cloth being sun-dried between each dip. Materials thus mordanted were then allowed to cure for a year before being immersed in dyes.

The type of ash to be used was selected with the intended dye in mind — different plants contain different combinations of elements and so ash from different species will also have a bearing on the eventual colour outcome.

While playing with the contents of the refrigerator one day I discovered that cow's milk does just as well, if not better. I experimented by painting my cloth with a pattern using milk and a brush, allowing it to dry and then dyeing. The areas where the milk had been painted were a much richer and deeper colour than the rest of the cloth.

Applying various layers can be a great deal of fun and can contribute beautifully to the colours you will eventually achieve. Here's what I would do with, for example, a reconstructed dress made predominantly from linen, hemp or cotton but with a few silk embellishments. Firstly I would plan a visit to the seaside. (If it's nice weather, take a picnic.) Providing the garment has been well rinsed after the last wash there shouldn't be any harm to the local ecology in either floating it in the waves or letting it soak in a rockpool. Dry it in the sun afterwards, being careful to cover up the silk bits (wet silk has a nasty habit of disintegrating when exposed to strong sunshine).

When the garment is dry, take it home and give it a swill in a dairy rinsing mix or soy protein. After the protein dip, let it dry in the shade. Next you could mix half a teaspoon of ash from the fire with a bucket of warm water and give your garment a brief baptism in the brew. Dry again before giving it a final immersion in more of whatever protein solution you have on hand. Once dry, let the garment simply hang for a week before tackling the dye-pot. When making multiple applications only the first of the brews should be applied in a long marination and all of the others should merely use a dip-and-pull technique. A series of long soaks might undermine the process by loosening parts of the earlier deposits.

An easy method of collecting a protein mordant

When rinsing milk, cream or yogurt containers for recycling simply pour the first vigorous rinsing into a receptacle kept in the freezer. Yogurt containers are splendid for this purpose. Add layers over time until the container is filled. When you are ready to apply mordant to a garment, thaw the container full of protein and water. A one-kilogram (two-pound) capacity yogurt container full of dairy-flavoured rinsing water is enough to mordant a large linen shirt or shift dress. You can also store the runny stuff that leaks out of tofu in the same way. Combine the two, if you like.

Of course you can vary the layers and the amusements as you go. A couple of years ago I celebrated the last evening of a visit to San Francisco by taking a customised silk thrift-store dress down to the beach below Ghirardelli Square. I'd spent a week happily running red lights on memory lane and so thought it would be appropriate to mark the time by collecting a mnemonic layer on the cloth. I stood there by the bay wearing my habitual 'r(eco)fashioned' garments and cowboy boots, hair dangling down my back in a long braid, and watched the dress flop about in the wavelets while gazing at the Golden Gate Bridge in the sunset. It was a pleasant activity and time passed. When I recalled myself to the present and retrieved the dress I realised that I'd been the subject of some attention. More than a few tourists had been filming the 'ceremony'. Maybe anthropologists of the future will wonder about the curious supposed Native American custom of soaking clothes in the sea. It afforded me no little amusement. After removing the object from the water I took it back to the apartment I'd been staying in, where I wrapped it firmly around a collection of iron railway-side debris that I had collected in San José a few days earlier. Overnight, lovely blue and purple stains appeared on the cloth where the iron pieces reacted with the residual moisture from the seawater.

left; railway tracks are a good place to look for scrap iron; right: treasure harvested from the railway-side

detail of dyed silk dress. I called it 'yerba buena', which translates as 'good herb' and was the Spanish name given to the city now called San Francisco ▶

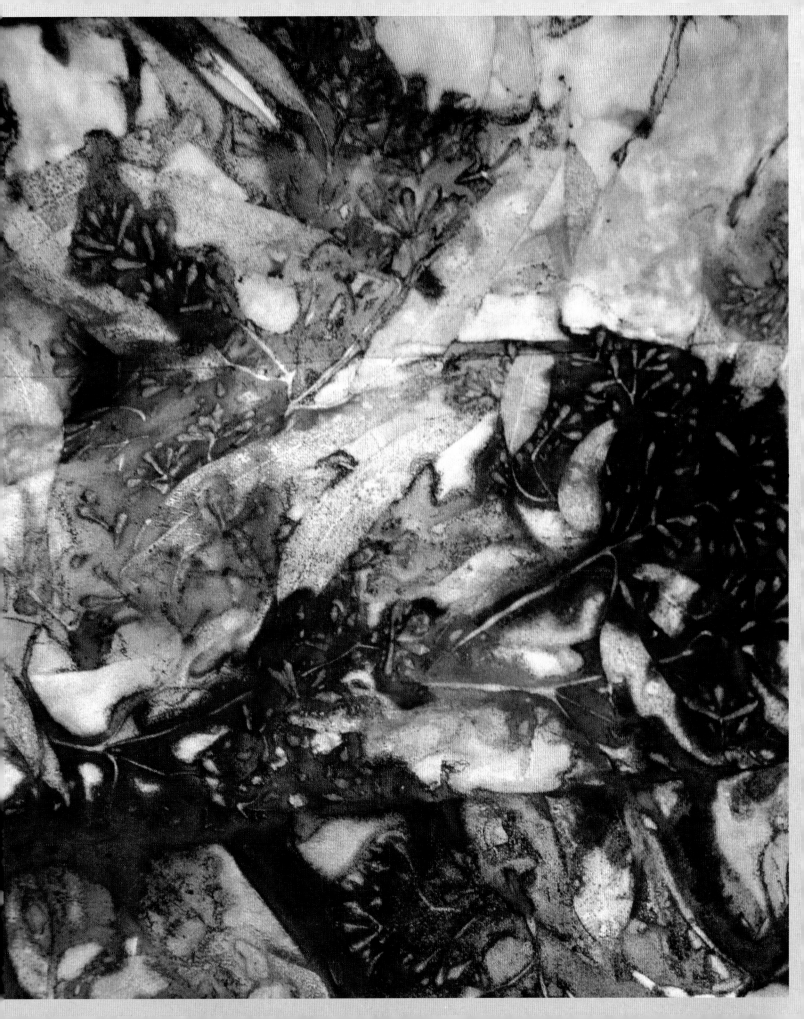

A month later and back home in Australia I dyed the dress using the eco-print method. Windfall leaves from a eucalyptus that usually prints a rich red gave me lovely purples, blacks, greys and even hints of lemon and pink as a result of my bayside ritual.

If your recycled garment is made of silk or wool you don't really need to do anything to it before dyeing as it will pick up some sort of colour no matter what. However, if you want lovely dark sinister shades, consider soaking it in a solution made up of a cup of iron water (see instructions on page 257) mixed with half a bucket of tap water. Let it rest in the mix overnight and allow it to dry before dyeing … or dye it while still damp for watercolour surface effects.

Things worth the knowing

The only real rule that I think must always apply in the use of plant dyes is that of knowing what you are dealing with. Every plant has the potential to give some kind of colour in the dye-bath but you need to be able to identify it accurately to decide on how best to make use of it. Some plants lose their colour if too much heat is applied, others don't release their optimum colour unless heat is applied. More importantly many plants are extremely poisonous and to innocently use hazardous substances is just asking for trouble. Folk sayings such as the North American 'leaves of three, let it be', referring to noxious irritants such as poison ivy and poison oak, were coined for good reasons!

eco-print from the leaf of the teak tree (*Tectona grandis*)

When collecting in the wild it is important to be able to accurately identify the plant, as harvesting a rare or protected species could seriously affect its long-term survival. Wild harvesting is a chancy thing in any event, even if you can identify the plant, decide it is acceptable to pick and only take a maximum of 10 per cent of the population; when your friend Betty sees the lovely colour and discovers how you made it and harvests her 10 per cent of the (remnant) population and then Bob down the road gets into the game – well, you can see where this is going. Before you know where you are you'll be in a similar pickle as the Phoenicians, sitting there by the side of the Mediterranean Sea surrounded by stinking piles of dead shellfish[41] and wondering how on earth they were going to satisfy the demand for purple togas. Luckily for them lichen dyes emerged as a substitute (and as a bonus smelled better), but we can't all count on luck to pull us out of the quicksand, and even then the substitute itself came under threat as a result of demand. Humankind has made rather a habit of trusting to fortune and as a consequence we find ourselves on the brink of the abyss. We've not done very kindly by Mother Earth.

I always look in my garden before going hunting elsewhere and when I do go into the wild I usually restrict my gatherings to windfalls. Canadian dyer and weaver Karen Diadick Casselman calls this practice 'salvage botany'. Whatever name you choose to give it, the act of collecting plant material that has already been separated from its growing location by some other action (whether Mother Nature or the roadside tree pruners) is a splendid way of making use of what would otherwise simply become compost (not that compost isn't important in the scheme of things). Even when collecting such pennies from heaven it's still important to be able to accurately identify the plant. Working in India a few years ago I used the windfall leaf from a teak (*Tectona grandis*) tree, bundling it tightly in a small silk bag in the hope of a print. I was indeed blessed with a lovely red print but unfortunately also bloomed a spectacular rash all the way up my arm from the tiny hairs on the leaf. I had been aware that the teak tree is considered poisonous and that teak wood shouldn't be used for toothpicks but it had never occurred to me that I'd have such a violent reaction to the leaf. The universe may sometimes offer painful lessons but at least they are not easily forgotten.

Onionskins

The easiest and cheapest dye-stuffs to begin your experiments with are onionskins (sometimes known in the vegetable industry as 'shells'). Keep a big brown paper bag in the kitchen and collect up all the skins from cooking preparations until you are ready to begin. It makes sense to keep the red (Spanish onion) skins separate from the brown as they each make different colours. If you're lucky enough to source pink-skinned garlic then keep those skins safe as well. Skins from the genus *Allium* will dye both cellulose fibres and protein fibres, meaning you can dye wool and cotton in the same dye-bath. Don't expect exactly the same shades though, as wool and cotton will take up dyes in different ways. In addition to making up a lovely dye the onionskins can be used to make contact prints by bundling them tightly in the garment or cloth and then boiling or steaming the bundle. Boiling the bundle will result in a nice *shibori-zome* pattern where the string you have used in the tying acts as a resist; on the other hand, steaming will ensure that no pesky string marks will compete with the prints. It's purely a matter of personal preference. When used in conjunction with scrap iron and/or copper, onionskins can often successfully dye synthetic fabrics as well.

an abundance of onionskins (or onion shells, if you prefer)

As onionskins are a by-product of cooking and an integral part of an edible plant, a dye made from them is perfect to use when working together with children as it does not affect their wellbeing if splashed onto the skin (unless of course it is boiling) and the skins can be arranged by them on cloth and even chewed on without unpleasant reactions. The safe brown dye from onionskins can even be used to add colour to gravy or stock if it's looking a trifle pale.[40]

red and brown onionskins on silk (the black is from an iron bar) ▶

Note, though, that not all foods are sourced from entirely edible plants – for example, potato and tomato leaves are poisonous, most parts of the peach tree contain cyanide and even something as harmless as apple seeds can be toxic in large quantities. I source large quantities of onionskins from wholesale vegetable dealers who are only too pleased to have them taken away.

Applications

There are many ways in which onionskin colour can be applied to cloth. For 'flat' colour, stuff the onionskins into a bag made of butter muslin or an old stocking to form a large 'teabag'. Immerse this 'teabag' in a sauccpan full of warm water and steep for a couple of hours to extract the colour. Boiling will release colour more quickly but steeping somehow gives a richer brew.

For those with plenty of time on their hands, the onionskins can also be processed using solar energy. Simply pack the onionskins into a large glass jar, fill with water, seal tightly and place the filled jar in a warm sunny place for a week or so. When the colour of the water has reached a satisfying depth, strain it off into another vessel for use. Refill the jar with more water and return it to its warm spot for further solar steeping.

Garments can be warmed in the rich brown solution in a saucepan or packed into a large glass jar and processed using the heat of the sun. If the latter option is chosen then variable patterns are unavoidable, so if flat colour is the desired outcome then stir your garment gently in a saucepan so that the dye can be distributed evenly.

Patterning

Should you have a preference for pattern, onionskins can be laid directly on the garment and the whole thing rolled and bundled tightly before being steamed, simmered, solar heated or even buried in the compost heap. The latter takes the most time and the outcome is best when attempted using resilient fibres such as linen, hemp and ramie. Wool and silk tend to be quickly devoured by micro-organisms living in such natural environments as compost heaps and gardens, so stick to the other options instead.

Variations

Lovely colour variations can be achieved through the addition of found metals to the dye-bath or dye bundle. Aluminium tends to encourage golds, copper can induce gold-green shades and iron deepens colour substantially. Depending on the water quality that is used for the substrate of the dye-bath (the table below assumes neutral pH and no dissolved salts), a few guides to possible outcomes follow.

	Brown onionskins	Red (Spanish) onionskins	Remarks
Iron	Yellow green	Olive green	The longer the dye solution remains in contact with the vessel the more intense the colour will become. Over a period of days in an aluminium pot an onionskin brew will become quite yellow; however, in an iron pot the brew will eventually darken to black
Copper	Golden brown	Copper brown	
Aluminium	Golden yellow	Chocolate	
Stainless steel/enamel (or other non-reactive vessel)	Warm brown	Dark brown	In a non-reactive vessel the water forming the substrate of the dye will have the greatest bearing on the colour outcome

onionskins on hemp cloth in the presence of iron

slow cold-dyeing using red onionskins and an aluminium tray with a little seawater

Magical auxiliaries from nature

The more I work with dyes from plants the more I find there is to discover. On another 'mordanting' excursion to the beach (this time in New Zealand) I found to my surprise that the colour on an organic cotton T-shirt recently dyed with onion and scrap metal was substantially enhanced when I casually tossed it into the brass planter pot in which other cotton garments were absorbing the benefits of a marine bath. Interestingly some of the colour was leached out of the shirt by the seawater and redistributed to the other garments but despite this, when the original was dried the onion colours were far richer than they had been before.

Such post-dye enhancement has often been referred to (and I must admit I have been guilty of this myself) as post-mordanting, but given that strictly speaking a mordant is 'a substance that creates a bridge or a bond between the fibre and the dye', this use of the term is not technically correct. Perhaps it might be more fitting to refer to the adjunct substances we use in our dye processes as auxiliaries. When they are used prior to dyeing we can call them mordants. When they are used simultaneously with and/or after the dye has been applied they would be more properly known as modifiers.

water from the Waikanae River (New Zealand) has substantially brightened the colour of the eucalyptus dye

An amusing means of modifying colours to be had from plants is to make up a number of solutions using a range of easily sourced liquids together with a range of found adjuncts. The beauty of this practice is that it rewards ingenuity by being largely free of cost (other than travel and that of a few of the purloined household ingredients) and substantially broadens the delight to be had from bundle dyeing.

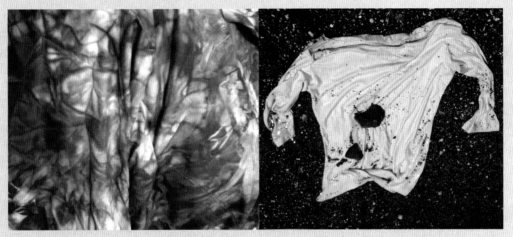

watercolour effect on silk using scrap iron and leaves in a bundle

organic cotton T-shirt soaking in seawater on a pebble beach in New Zealand

Making your own mordants is really very simple. Place the substance (scrap iron, copper or brass, aluminium foil, ash, crushed shells, rhubarb leaves or citrus peel) in a large clear glass jar. Use glass because it is non-reactive and also transparent, allowing you to see what's happening inside. Fill the jar with the liquid of your choice (tapwater, rainwater, borewater, seawater, urine, vinegar or exhausted dye-bath). Label the jar and let it sit for a month or two. There's no need to be limited by the list above – but avoid using things that are known to be dangerously poisonous, such as lead and arsenic. Enthusiastic brewers will also perceive that some of the solutions might lend themselves to receiving further additions … for instance you might add copper scrap to ashwater to see how the alkaline solution reacts with the metal. Without the variations in ash compositions (different plant species produce chemically different residues when burned) and exhausted dye-baths there are already at least 56 different mordant combinations possible from the substances listed. Small quantities of the solutions can be used diluted as pre-mordants, added to dye-baths or applied as a kind of post-dyeing colour modifier/developer.

Be sure to store these mixtures in some out-of-the-way place well out of reach of children and pets.

These relatively harmless solutions can be safely combined, although sometimes you'll see interesting reactions; for example, an acid solution introduced to a salt bath may cause a cloudy suspension to form. The mordants can also be diluted slightly, strained and then sprayed onto cloth using a hand-pumped spray bottle. Allow the cloth as much time as you can for it to cure and then proceed to dye in whatever way you wish. It would be advisable to wear a mask when spraying so as to avoid inhaling the mist; common sense suggests that even though the substances are not in themselves aggressive it would be wise to expose the body to the least harm at all times.

Eucalyptus

The magical combination of eucalyptus and wool continues to be woefully underestimated by the textile industry. There is a clear potential for leaf matter (an abundant by-product of both forestry and floristry industries) to be gathered and applied to wool not only for the beautiful colour but also for the scent that is imparted to the fibre. That eucalyptus dyes completely coat the fibres if applied properly has the added bonus of making the fibres more resistant to felting and therefore enhancing the longevity of a garment without using other toxic surface treatments. Bringing together these two Australian icons – the sheep and the gum tree – could have substantial long-term benefits for the local textile and fashion industry.

The important thing to remember about eucalyptus dyes is that the best and brightest colours are achieved in baths of neutral or slightly acidic pH. This is something that has not been properly understood even by dyers of long experience; more importantly, the addition of an acid such as vinegar to an after-bath for eucalyptus-dyed cloth *will not* act as a post-dyeing brightener. Eucalyptus dyes may possibly be enriched after their application with substances such as urine or vinegar-and-copper solutions, or darkened by applying iron-water, but if the original colour is a murky brown it will simply become an enhanced murky brown.

If you suspect the water you intend using to be alkaline or of having dissolved impurities then it is important to add an acidic solution to that water prior to immersing the leaves and beginning to extract the dye. A cupful of vinegar or half a dozen fresh rhubarb leaves (and their stalks) added to 10–20 litres (4–8 gallons) of water in the dye-pot and warmed slightly before putting in the eucalyptus leaves will ensure you get the best possible colour from the process. Stuffing the rhubarb leaves into an old sock or stocking can be helpful as the fibrous matter can sometimes prove tricky to remove from your garment later.

Interestingly the phenomena observed in a eucalyptus dye-bath can tell you a lot about the water you are using. Alkaline water will result in a murky dye whereas if the water contains dissolved salts, these will (in the presence of eucalyptus) manifest their presence as a cloudy suspension that hovers near to the bottom of the vessel. If you find, for example, that *Eucalyptus nicholii* is yielding purplish tones in a stainless-steel dye-pot you could hazard an educated guess that the water is not only acidic but also being delivered by iron piping.

Testing for pH

The very simplest way to test reticulated water for pH is to take a shower in it. Neutral to slightly acidic water will feel slippery on your skin, whereas alkaline (and salt) water will make you feel slightly itchy. This is one of the reasons I like to spend a night in a region before I teach a class there, as arriving fresh from the airport and stumbling into a morning class leaves no opportunity to familiarise oneself with the most important ingredient in the dye-bath – the local water.

Those with more time may like to take a reactive plant material (such as red cabbage, raspberries, elderberries or blackberries) and let it soak in a little methylated spirits for a few days. Once the spirits have absorbed good colour, soak cotton buds in the solution. Let them dry then store them in a glass jar. Dip into waters to test for acidity (pinky red) or alkalinity (blue). Alternatively, pH test strips are available from most chemists.

Whichever way you decide to apply colour to your garment, the mantra 'time is your friend' holds true. Colour in cloth that is dyed slowly, given time to absorb colour gently, and allowed to steep and then cool in the dye-pot will be far richer and much more durable than that which has been rapidly boiled, removed from the bath while still hot and then vigorously rinsed. The structure of wool in particular needs heat and moisture for good dye take-up; slow dyeing will reward you with well-bonded colour. Wool yarn steeped in a hot freshly prepared eucalyptus dye-bath, allowed to cool in the solution overnight and then dried (after a gentle machine spin to remove excess moisture) will then rinse clear, showing no colour leaching at all.

even synthetic fabrics can be dyed when crushed into bundles around scraps of iron

an exquisite eucalyptus and wool sample dyed by Marina Lamsens (Belgium)

Roll up and dye

If garments are to be processed in a very hot dye-bath it is important that they not be boiled *unless* they have been immobilised by bundling. Cloth that is allowed to bubble and boil will inevitably be affected by the process, leading to shrinkage in the case of wool or possible brittleness in the case of silk. On the other hand, garments that have been bundled together with leaves around solid objects and then firmly tied can be safely boiled. While this of course means that colour from the pot is only absorbed by the exposed parts it also means that colour from the plant material in the bundle can be transferred directly to the cloth without being diluted by water. It is a technique that makes for very exciting possibilities.

bundle-dyed patterned silk cloth ▶

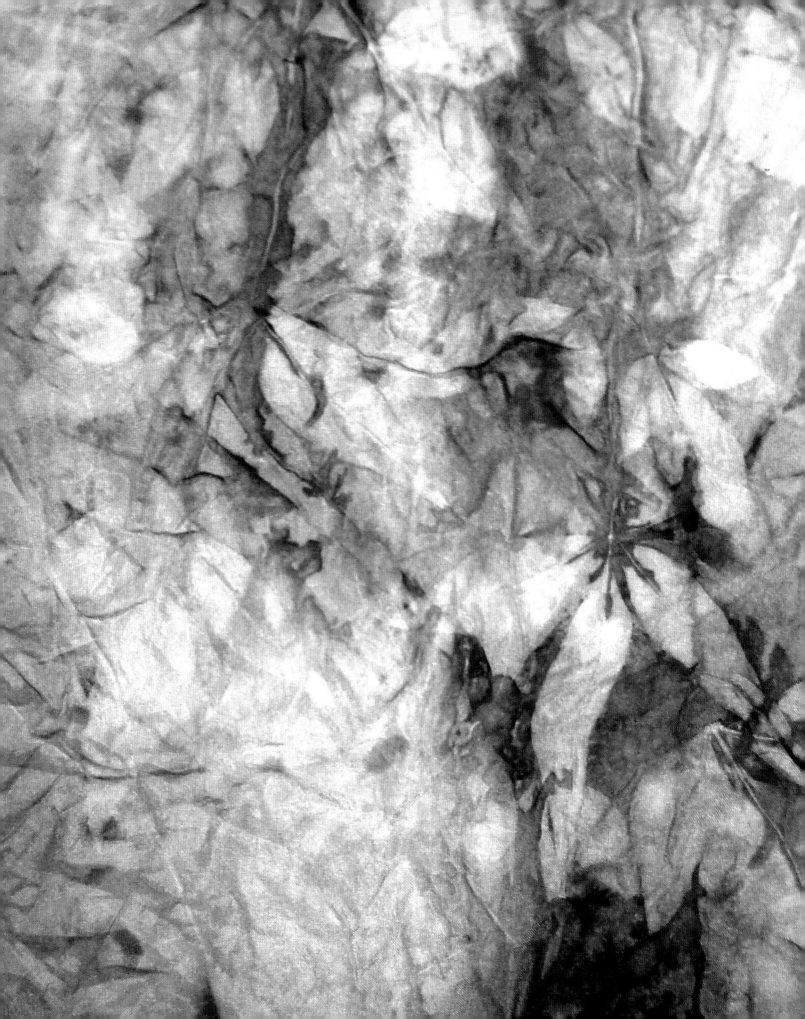

The windfall walk

I am blessed with a somewhat peripatetic life and whenever I arrive in a new place, while out and about in the world I like to take what I describe as a 'windfall walk'. Essentially it encompasses taking a walk and collecting windfall plant material, but the point of difference is that I preset the points at which I will stop and collect material rather than actively searching it out. This may seem a little strange but it is important when wandering in unfamiliar territory, as you need to be able to remember a few landmarks and directional clues to get back to wherever you might be staying – and if you're walking about with your nose firmly pointed at the ground looking for pretty leaves it is very easy to become hopelessly lost, unless you are very good at remembering pavement patterns or are following a yellow brick road. So I choose arbitrary stopping points such as reciting poetry to myself or singing Broadway hits aloud and stopping at the end of each verse (such behaviour has the added benefit of convincing the populace that one is mad and therefore to be avoided) or I might stop at the sound of a car horn (not a good one to use in India, they never stop tooting) or if I hear a dog bark. Whatever the chosen stopping point, I look down at the ground and pick up the leaf that is nearest my feet, deposit it in the capacious pocket of my apron and walk on. Of course if I see a particularly interesting plant in my wandering I'll walk to it and purposefully collect specific offerings from the ground beneath it as well.

the stairways of San Francisco are always good for a leaf or three ...

clockwise from top left: eucalyptus eco-print on hand-woven Indian silk shawl; in my garden with studio assistant Martha; windfall walk notes; ▶
eco-prints on velvet using leaves gathered from the the farm where I live (the darkness is due to wrapping the cloth around an iron bar);
a collection of windfalls gathered in a forest in India

So far, this 'windfall walking' method has resulted in pleasant perambulations around interesting places without losing myself. Eventually I head back to the hotel or wherever I might be staying and then spend some pleasant minutes laying my treasures out on a piece of silk or wool cloth, rolling it up tightly and tying it firmly with a piece of string. I then unpack my tiny travelling dye cauldron, essentially a small electrically powered die-cast aluminium vessel, a lucky find in a country opportunity shop in Victoria.

I put the windfall bundle in the dye-pot along with any spare leaves or flowers I may have picked up, adding used teabags for luck, plug it in near an open window or in the bathroom so as not to set off the fire alarms, cover the bundle with water and simmer it for about half an hour. Then I turn the pot off and retire for the evening. In the morning, prior to any teaching or research commitments I enjoy a delightful start to the day by opening my windfall bundle. What I find inside offers a wealth of information about local plants and the tap water and sometimes even reveals a new and quite unexpected result.

When teaching I adjust the windfall walk slightly so as to place a time limit on the wandering, otherwise I might find students taking off for the hills and not coming back to class until teatime; it's such an absorbing activity. Students are given a length of silk or wool cloth about a hand's breadth in width and no longer than the height of their shoulders from the ground (any longer and it's too easy to trip). They begin by selecting a nice firm stick, not too long, or if we're near a river then a nicely shaped stone. Choosing their own stopping points they begin to walk, wrapping each leaf into the cloth (and around the stick) as they go. When the cloth is full they all return to the starting point, tie their bundles tightly and proceed with the dyeing.

Disposing of dye-baths

I have refined these dye methods on principles of least harm to both the dyer and to the ecology. But they aren't completely harmless – for example, you wouldn't want Fido to drink brews made from St John's wort or your toddler to be splashing about in a pot laced with urine. I dispose of spent dye solutions by pouring them on my compost heap. From time to time (whenever I clean the fireplaces) I sweeten the heap with a little ash (the rest is dug into the vegetable patch). Even dye-baths that have been spiced with seawater mordants may be disposed of in this way as the quantity of salt remaining will be minimal (the active elements will have bonded to your cloth with the dyes). If you live in a city apartment where a compost heap may not be practicable, you may safely dispose of solutions using the common drain.

Each person tailors their walk to their own inclination or capability. On a memorable occasion in New Zealand a friend of mine in the class who had delivered her baby by Caesarean section only five weeks before and wasn't yet up to stomping about the hills and dales limited her walk to the inner courtyard of the campus. As it happened, this made our search for the plant that had given a miraculous pink on her cloth very much easier – Rachelle Toimata had made the discovery (hitherto undocumented in dye literature, so far as I have been able to ascertain) that a yellowed and slightly withered leaf of *Griselinia littoralis* could print a rich, clear pink. Further investigation of this genus confirms that *Griselinia lucida* does the same and that fresher, greener leaves give an even brighter colour. The funny thing was that I had been visiting that campus and teaching there for a couple of years, and until Rachelle took her 'courtyard windfall walk' no-one (me included) had bothered to try the leaves of this plant, even though it was growing right outside the classroom door. Even unlikely plants can offer amazing colour.

far left: *Griselinia littoralis* eco-print; left: *Griselinia littoralis*, a plant with surprise

A couple of years later, while teaching in the Tennessee woods, my students and I discovered that the sourwood tree (*Oxydendrum arboreum*) yields blue prints from its tiny white flowers as well as from the mature leaves.

Bundle dyeing, whether random or deliberate, offers a wealth of possibility. More colour and pattern variations can be introduced by adding fragments of found metals such as flattened rusted bottle caps, found paper clips and old nails or by wrapping the bundle around larger pieces of found metal. When sticks are used the bark will often make an interesting print on the inside of the roll. The technique of 'pot as mordant'[41] can have a strong influence as well, particularly if there are plant parts included in the cooking solution.

It can be a very pleasant and meditative occupation to hand make (or for that matter, rehabilitate) a garment, then to spread it out on a flat surface and lay down leaves, flowers and bark together with other interesting finds such as small rusted objects or 'drawings' formed from wire, then gradually fold the whole inward until a firm parcel has been achieved. Tying the bundle firmly with string, finishing with two half-hitches so that the knot can be easily untied (and the string reused) and placing it in a suitable pot to simmer for a while, leaving it to cool overnight and then preferably to rest for a couple of days is sometimes an interesting test of patience but will be well rewarded. Instantly opened bundles, although beautiful in themselves, do not always realise their full potential. This unhurried approach – taking time in the making and sewing, in the mindful gathering and gentle low-impact dyeing, by which maximum colour is coaxed directly from a relatively small volume of plant material – leads to garments that are far more valuable than their materials alone. These are second skins that have been made with reverence and thought and commitment, with consideration for the ecology and for the other occupants of the planet: clothes that will be worn and washed with care, patched and mended and worn again. It is a style of dressing that requires a bit of backbone, for someone not easily swayed by fashion or by the desire to conform or worried by thoughts of what others might think. It's how I like to wander through the world and I know I'm not alone – this book has been written for you.

workshops yield a surprising array of colour, even when the source is only florist waste

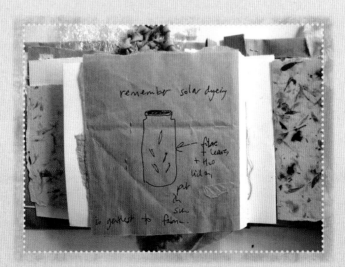

for those with time on their hands, solar dyeing is the most energy-efficient means of applying colour – bear in mind, though, that for brilliant colour from eucalyptus, the heat of boiling is a vital catalyst

chapter 11
And in the very end

wrap me up with my stockwhip and blanket

and bury me deep down below

where the dingoes and crows can't molest me

in the shade where the coolibahs grow

Andrew Barton Paterson, 'The Stockman's Lament'

The fairly recent fashion for extravagant burials and cremations complete with elaborate caskets seems a huge waste of resources to me. In the past, the poor were often buried naked (clothes were too valuable to waste) or in a winding cloth or simple shroud, the latter often stitched together from several pieces of cloth. In Britain the Parliament passed the Burial in Woollen Acts 1666–80 prescribing that bodies were to be buried either wrapped in wool or naked:

No corpse of any person shall be buried in any shirt, shift, sheet or shroud or anything whatsoever, made or mingled with flax, hemp, silk, hair, gold or silver, or in any stuff or thing other than what is made of sheeps wool only.[42]

While it is likely this was a protectionist move designed to promote the use of British wool it does seem rather more sensible than burying a perfectly good suit of clothes or chopping down an oak tree to construct a coffin. How much more beautiful would it be, to be quietly interred in a vertically drilled hole in a memorial forest, wrapped in a simple piece of cloth. The body would provide nourishment for a tree planted in the nicely loosened earth and could continue to contribute to the world in a really useful way.

A shroud made of thick, naturally coloured felt would provide a suitably sturdy vessel for bringing the body to the burial field. Creating the felted shroud could be an event that involves the bereaved, offering comfort in the vigorous activity of rolling, perhaps followed by some gentle stitching time adding messages or symbols of significance to the cloth. Occupying the hands with such simple tasks leaves the mind the necessary space to grieve as well as being a way of paying one's respects to the dead. The woollen shroud would not only keep the body together during the process of interment, it would also act as a slow-release nitrogen-rich fertiliser to help nourish the trees of the memorial forest.

Making a quilt from some of the clothes of the deceased could also offer a comforting occupation for the bereaved. Sewing circles are wonderfully therapeutic and stitching a beautiful textile from a selection of favourite garments belonging to the loved one just passed creates a useful and practical object rich in stories. Consider stitching text onto the quilt, recording particular memories or favourite phrases and turns of speech. One might simply build up layers as the years pass, adding patches and remembrances here and there as family members inevitably pass on; a narrative-rich memento to hand on through generations.

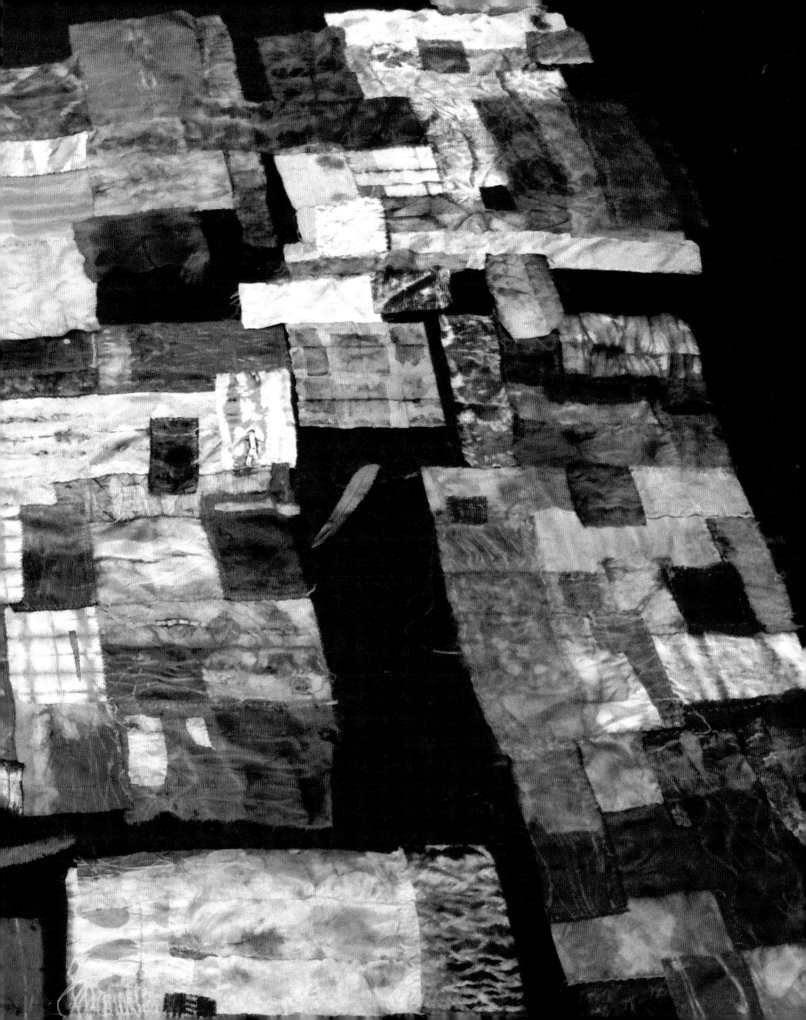

Notes

1. Gwen Egg, fibre and textile artist, back cover of the exhibition catalogue 'Returning', published by Clarence City Council, Tasmania, 2009.
2. J R R Tolkien, *The Fellowship of the Ring*, Unwin Paperbacks, 1985.
3. http://www.ifm.eng.cam.ac.uk/sustainability/projects/massuk_textiles.pdf 'Well dressed? The present and future sustainability of clothing and textiles in the United Kingdom'.
4. L Pardi, in 'Social Fabric', *Peppermint* magazine, Issue 6, p 68.
5. Colin MacDowell is fashion writer of the *Sunday Times'* Style Guide and author of numerous books on fashion.
6. http://www.asbci.co.uk/conferencesbrief.php, 17 June 2007.
7. E W Barber, *Women's Work – the First 20,000 Years*, WW Norton & Company, New York & London, 1994, p 43.
8. G Naylor, *William Morris by Himself – Designs and Writings*, Time Warner Books, 2004, p113.
9. The appellation 'organic' is used here to describe substances that originate from 'living things', rather than implying that their production is untainted.
10. Slavery is still rife in the modern world, with numbers (reputedly more than 12 million) at their highest level in history. India has more slaves than any other country but the practice is prevalent on every continent.
11. New Zealand Wool Board, 1974.
12. Mary Grant Bruce, *Back to Billabong*, Ward, Lock & Co Ltd, 1921.
13. http://www.secondnatureuk.com.
14. Exodus 26: 7 in the King James Bible states: 'And thou shalt make curtains of goats' hair to be a covering upon the tabernacle: eleven curtains shalt thou make' and goes on to give strict instructions as to size, construction and installation.
15. Robert R Franck (ed), *Silk, Mohair, Cashmere and Other Luxury Fibres*, Boca Raton: CRC; Cambridge: Woodhead, 2001.
16. Georg Von Georgievic, *The Chemical Technology of Textile Fibres – Their Origin, Structure …* , Scott, Greenwood & Co, London, 1902, p 180.
17. K Fletcher, *Sustainable Fashion and Textiles*, Earthscan, London, 2008, p 166.
18. S Yanagi, *The Unknown Craftsman*, Kodansha International, Tokyo, 1989, p 117.
19. M Braungart and W McDonough, 2002, *Cradle to Cradle*, North Point Press, p 59.
20. I admit guilt here myself, having named my dye book *Eco Colour* in 2008.
21. Unless they are completely dissolved in chemicals, in which case they can be reformed as filaments – but that brings the added problem of managing the chemicals and how they may be safely disposed of.
22. A Allwood et al, *Well Dressed? The Present and Future Sustainability of Clothing and Textiles in the United Kingdom*, University of Cambridge Institute for Manufacturing, 2006, p 68.
23. Barbara Vinken, *Fashion Zeitgeist – Trends and Cycles in the Fashion System*, Berg, Oxford & New York, 2005, p 69.

24 TIME magazine, Coco Chanel obituary, 25 January 1971, author unknown: http://www.time.com/time/magazine/article/0,9171,904672-2,00.html#ixzz0n6uf3Jyx.

25 See Chapter 10 on dyeing with plants for an explanation of this term.

26 The stone was later returned to the river.

27 Henry David Thoreau, *Walden* (Chapter 2, 'Where I lived and what I lived for'), Penguin Books, first published 1854, this extract 2005, p 27.

28 A Allwood et al, p 68.

29 M Attwood, *Alias Grace*, Virago, London, 1997.

30 Sharon Sadako Takeda and Luke Roberts, *Japanese Fishermen's Coats from Awaji Island*, UCLA Fowler Museum of Cultural History, Los Angeles, 2001.

31 The new owners of Project Alabama moved the workbase to India and the company no longer has any links with its roots beyond the name.

32 www.denhamthejeanmaker.com.

33 Quoted in P Arnett and E W Metcalf (eds), *Mary Lee Bendolph, Gee's Bend Quilts and Beyond*, Tinwood Books, Atlanta, 2006, p 14.

34 http://www.resurgence.co.uk/magazine/article2459-guerrilla-bagging.html.

35 http://www.morsbags.com/.

36 Gerald Durrell, *My Family and Other Animals*, Penguin, 1956.

37 A matted textile described as a pre-felt or semi-felt is strong enough to hold together but fibres can still be pulled from the surface.

38 Sometimes stains will act as a mordant and react dramatically with the dye-bath. I once overdyed a cream coloured jumper that had no visible marks on the surface only to discover when I pulled it from the pot that there were spectacular black spots down the sleeves. So I set the spots as the centre of daisies and embroidered happily all around them.

39 The practice of harvesting shellfish for purple continues to this day. It involves the removal of the hypobranchial gland from the animal and the squeezing of its contents onto cloth. In sunlight the yellow smear photo-develops to become purple. In Mexico they describe this somewhat romantically as 'the tears of the shellfish'. Well may they do so, seeing the poor creature loses its life in the process.

40 Assuming the onions have not been sprayed with a fungicide.

41 I had thought that the use of pot as mordant, whereby the metal of the pot is selected for its potential effect on the colour outcome, was a relatively recent development in the plant dye world (although it must have been a consideration in early history) as for years the recommended practice had been to employ stainless steel or enamel pans for the preparation of dye. However on a visit to Te Papa museum in Wellington, New Zealand, I was excited to be shown the dye samples made by local dyer Katharine Phillips in the 1950s. Her notes indicated that she too was very familiar with the use of dye-pot as mordant and moreover spelled out her concern that natural dyers were using toxic mordants in excessive quantities. Her opinion seemed to be that such substances should be used in moderation, if at all.

42 The Burial in Wool Acts 1678. Source http://www.british-history.ac.uk/report.aspx?compid=47386.

Dancing with the interpixies (useful websites)

sustainable living

stories about the origins of various textiles and their impact on the world
http://oecotextiles.wordpress.com

a website about the provenance of fibres for clothing
http://www.swicofil.com/natural.html

the sustainable cotton project
http://www.sustainablecotton.org

An Education Lab for Socially Responsible Fashion Design
http://socialalterations.com/

The Social Network for Sustainability
http://www.wiserearth.org

a plan for repurposing things
http://www.zerolandfill.net/

hemp

online hemp store stocking fabrics and bedlinen
http://www.hempgallery.com.au/

silk

online silk store stocking fabrics and threads
www.beautifulsilks.com

wild silk

an ecologically sustainable fibre project in the Himalaya region
http://www.avani-kumaon.org

nettles

About clothing from nettles
http://www.nettles.org.uk/nettles/people/clothing.asp

wild fibres

spinning and weaving with natural fibres from nettles to cashmere
http://www.wildfibres.co.uk

A place where anyone can learn more about the valuable and precious world of natural fibres by sharing knowledge, skills, resources and products
http://www.keepthefleece.org

Wild fibers magazine
http://www.wildfibersmagazine.com

clothing

Denham the Jeanmaker
www.denhamthejeanmaker.com

Sugarcane Jeans
http://www.sugarcanejeans.co.uk/

the Little Brown Dress Project
www.littlebrowndress.com/

a non-for-profit association that helps to sustain handcraft knowledge

http://www.edelkoort.com/projects_heartwear
http://www.heartwear.org/

Fashioning Now: changing the way we make and use clothes

A project from the University of Technology, Sydney
www.fashioningnow.com

individual clothing by Emma Christie
www.emmajanechristie.com/

sewing tips and tricks

http://sewingneedlework.suite101.com

interesting website with lots of useful links
http://www.sew-green.org/

free patterns

http://www.freeneedle.com/ is an online compendium of free sewing patterns

http://www.smockshop.org/ has a downloadable smock pattern

wool

Milkymerino, a new way of processing wool
www.milkymerino.com

making the most of naturally coloured wool
http://www.chocolatewoolnz.com/

artists

Anu Tuominen
www.anutuominen.fi

Holly Story
www.hollystory.com

Jude Hill
http://spiritcloth.typepad.com

Christine Mauersberger
http://cmauers.blogspot.com/

Museums

Museum of Economic Botany
Botanic Gardens and State Herbarium,
South Australia
http://www.fbga.asn.au/
MuseumEconomicBotany.htm

Pioneer Women's Hut, Tumbarumba
http://www.pioneerwomenshut.com.au/

Textile Museum, Toronto
http://www.textilemuseum.ca/

Amuse Museum, Tokyo
http://www.amusemuseum.com/english/facility/index.html

Migration Museum, Adelaide
http://www.history.sa.gov.au/

Paperchase (further reading)

Barber E J W, *Prehistoric Textiles*, Princeton University Press, 1991

Black S, *Eco-chic the Fashion Paradox*, Black Dog Publishing, London, 2008

Blanchard T, *Green is the New Black*, Hodder & Stoughton, 2007

Braungart M & Mc Donough W, *Cradle to Cradle*, North Point Press, 2002

Casselman K D, *Lichen Dyes: The New Source Book*, Dover, 2001

Colchester C, *The New Textiles – Trends and Traditions*, Thames & Hudson, London, 1991

Diamond J, *Collapse: How Societies Choose to Fail or Survive*, Penguin Allen Lane, 2005

Flint I, *Eco Colour*, Murdoch Books, Sydney, 2008

Harris J, *5000 Years of Textiles*, British Museum Press, 1993

Katoh A S, *Japan Country Living*, Tuttle Publishing, Vermont, 1993

Koide Y and Tsuzuki K, 2008, *Boro – Rags and Tatters from the Far North of Japan*, Aspect Corp., Tokyo, Japan

Ministry of Information, *Make do and Mend* (facsimile), Imperial War Museum, London, 2007

Sanders A & Seager K, *Junky Styling Wardrobe Surgery*, A & C Black, Britain, 2009

Yanagi S, *The Unknown Craftsman*, Kodansha International, 1989

Yoshida S and Williams D, *Riches from Rags*, San Francisco Craft Museum, 1994

Index

Page numbers in *italics* indicate illustrations.

A

Acacia 69
Acer palmatum (Japanese maple) 126, *127*
adire cloth (Africa) *157*
Alabama Chanin 166–7
Albert, Prince 106
alkalinity 218
Allium 252
alpaca 63
alum 246
aluminium 255, *255*
American Civil War 52
anaerobic digestion process, of fibre production 550
angora 63
anti-microbials 76
apple seeds 254
appliqué 166, 211
aprons 131
Aral Sea 53
asbestos 44, 45
ash, use in dyeing 246, 247, 257
Asia 90, 131
   cotton industry 91
      garment fragment *34*
Assisi, St Francis of *158*
Australia
   cotton 50, 52
   organic sheep 62
   textile industry 90–1
Avani Kumaon project 68–9
Azadirachta indica (neem) 76

B

bags
   from jeans 193
   from jumpers 195
   from pilchers 233, *233*
   plastic 194
   from singlets 204
   from skirts 193
bamboo (*Bambusa* sp.) 55
banana fibre 45, 53
banana peels 140
banksia *236*
Barber, Elizabeth Wayland 35
bargasse 69
bark 45, 266
bashofu cloth 53
Bates, Daisy 107–8
beads 197
beef cattle, and cotton trash 52
beer 46
bereavement 271
births 18
black
   dyeing wool 62
   fading 149
   sulphur black 72
   using logwood 73
   wearing of 100
blanket stitch *160*
bloodstains 152
Boehmeria nivea (ramie) 54, 254
Bombyx mori (silk moth) 65
Bonaparte, Napoleon 237
boro textiles 39, 120, 158–9, *161*
   floorcloth from Japan 224
bracelets 231
'Breath' video 172, *173*
Brennels 55
bricolage 182
'Bridal' dress *167*
Britain
   austerity 116, 136
   burial in wool 270
   kilts 106
   pollution 89
   synthetics 23
Bruce, Mary Grant 59
bundle dyeing 242–3, *242*, 252, 257, 260, *261*, 266
Burial in Woollen Acts (1666–80) 270
burials 270
burn test 118
button plackets, reuse 203
buttons *130*, 132–3, *197*
   to add interest 197, 201
   hand-made 184, *185*
   threading 233

C

Caldwell, Dorothy 178–*9*
calico 115
Cannabis sativa see hemp
carbon 22, 43
carbon storage 58
carcinogens 72, 237
cardigan, from jumper 200
carpet cleaning 141, 142
casein 45, 76
cashmere 63
Casselman, Karen Diadick 251
cellulose cloth 244, 245
   preparing to dye 246, 252
Chairman Mao 105–6
Chanel, Coco 102–3, 114
Chanin, Natalie 166–7
chemicals 43, 55, 85
   impact on land 52
   on natural fibres 87
   next to skin 29
child labour 28
China 35
   sericulture 65, 67
Christie, Emma 88, 164–5, *185*
chrome tanning 69
cleaning rags *18*, 211
cleaning techniques 136–7, 140–53
cloth drawings 176, *177*
clothes coupons 116
clothing 19, 21
   air-drying *136*, *146*, *154–5*
   airing 149, 153
   appropriate 80, 94
   choosing 86–7, 94–5
   contact with skin 29, 237
   'green' 28
   maintaining 92
   from nature 32
   quality 85
   ready-made 25
   repurposing 189–92, 195, 197
   sourced on internet 91–2
   washing 136–7
   see also garments
clothing production 23, 37, 89–90
   wastage 91
clothing workers 89, 90–1, 95
coal 22–3
coat
   from jumper 200
   reconstructed *198*
coffins 270
collars
   from jumper *198*
   reuse as accessory 203
colour choices 99–100

composite cloths 26
compost-heap dyeing 254
consumer goals 95
consumerism 25, 107
consumption 24–5, 108
contact prints 252
copper 255
copper-rich water 244–5
cotton (*Gossypium* sp.) 43, 50–3
  dyeing 252
  organic 53, *53*, 91
  shrinkage 119
  source 45
cotton industry 27–8, 52
  in Asia 91
crease resistance 76
crease test 118
cremations 270
crochet shapes 203
cuffs, reuse as accessory 203
cushion stuffing 211

D

darning 159, 160
  decorative 243
  hiding 243
  invisible 161
Denham the Jeanmaker 85, 168–9
denim 141, 201
dermatitis 72, 251
desertification 63
design copyright 121
detergent 146
DPK Fabrics 89–90
drawstring trousers 131
dress, cost of making 90–1
dress form 129
dresses
  hand-sewn patchwork *196*
  patched *188*
  plant-dyed *96*, *103*, *126*, *188*
  with string lace *230*
dressmaking
  choice of material 118–19, 122
  equipment 123, 128–9
  first attempts 115
  local outlets 90
  washing cloth 119–20
dry cleaning 143–4
dye cauldron, travelling 264, *264*

dye houses, and pollution 53
dye-baths, disposal 265
dyeing 236–47
  bundle dyeing 242–3, *242*, 252, 257, 260, *261*, 266
  effect of water 244, 255, *256*, 259
  mordants 245–6, 257–8
  with onionskins 252–5, *252*, *253*, *255*, 256
  post-dye enhancement 256
  time effect 260
  treatment for fading 243
  use of seawater 247, 248, 256, *257*
  using florist waste 267
dyes 72–5, 100
  home use 75
  sulphur black 72
  synthetic 53, 72, 118
  vegetable-derived 118
  yellow 237
  see also plant dyes

E

Earth, damage to 22, 251
Echium plantageneum 27
Eco Colour 241
eco-friendly labelling 87
eco-prints
  baby jumper *109*
  eucalyptus on wool *240*
  rose leaf *239*
  teak leaves *250*
  technique 118–19
  on velvet *263*
Edelkoort, Lidewij 105
Edwards, Trish *204*
Egypt
  cotton 50
  mummy cloths 54
  retting 47
emissions 22, 89
energy use 86, 87, 92, 118
environmental impacts 23, 25, 44
  cotton 52–3
Eucalyptus cinerea 109, *239*, *240*
eucalyptus dyes 58, 238–41, 244, 258–9
Eucalyptus nicholii 259
eucalyptus prints
  baby jumper *109*

dress *188*
Milkymerino 67
on silk 64

F

fabric conditioners 147
fabrics
  old with new 122
  plant-dyed 197
  sale of bolts 132
  see also materials; textiles
fashion 19, 24, 83, 102, 105
'Fashioning Now' (exhibition) *102*
fast fashion 25, 28
felt 34–5
  fulling 213
  pre-felt 213, 215, 275
  with silk scraps *210*
  from wool sliver *211*, 212–15
felting 119, 213, 217
  tools 218
feltquilt 227, *229*
flax (*Linum usitatissimum*) 45, 46
flax meal compresses 46
flea markets 132
Flinders Ranges 244
Flint, Violette 195, *196*, *201*, *204*, *233*
flour bags 82
footprint, on Earth 95
footwear 90, 204, *204*
fossil fuels 23
France 21
frogs 146
frugality 116, 189
fulling 213
fur 35

G

garage sales 132
garlic, pink-skinned 252
garments
  from crochet shapes 203
  final uses 211
  from rag strips 204
  reconstructing *204*, *205*
  refreshing 197
  from string 204
  unpicking 120–1

'Genes/Jean's' *181*
gloves, fingerless 195, 205
goat fibres 63
Gossypium sp. (cotton) 43, 45, 50–3
   G. arboreum 50
   G. barbadense 52
   G. herbaceum 50
   G. hirsutum 52
grandmother, author's
   clothes 114
   colour choices 113–14
   gift as child 13
   sewing machine 30, *30*, 112–13, 128
grease stains 153
greasy wool 140
Great Depression 82
Griselinia littoralis 265, *265*
Griselinia lucida 265
growth, economic 22
Gunter-Brown, Cory *198*

H
Habitat 83
Haematoxylon campechianum (logwood) 73, 75
hand sewing 123, 125
hand washing (clothes) 137, 144–7
harakeke 45, 47, *47*
hard water 136
Harford, Nancy 142
haute couture fashion 28, 91
Hawker, Roz 176–7
heavy metals 72, 76, 238
Helix (chlorfluazuron) 52
Hemp Act 1948 48
hemp (*Cannabis sativa*) 45, 48, *49*, 65, 254
   trousers *133*
Highland dress 106
Hildegard von Bingen 46
Hill, Jude 170–*1*
holes, in clothes 160–1, *160–1*
home-made clothes 25, 30, 31, 112–13
hot-bundle dyeing 242–3, 252, 260, *261*
Hypericum perforatum (St John's wort) 126

I
India
   'bejewelled' fabric 32
   cotton *51*
   flat iron *156*
   goats 63
   kantha cloths 39
   rafoogar 159
   sacred cow 24
   synthetic waste 71
   washing clothes *145*
indigenous Australians 107
indigo 119, 193, 244
   research program 74
individuality 24, 266
infant mortality 237
Ingerton, Roger 19
internet 107
   fashion news 103
   made-to-order clothes 91–2
iron, use in dyeing 248, *248*, 252, 255, *255*, *257*, *260*
iron water 250, 257
ironing 120, 156
   avoiding 156

J
Jackson, Vanda 159
Jamieson 126
Japan *140*
   bashofu cloth 53
   boro textiles 39, *224*
   ceramic fragment *29*
   dyeing techniques 246
   floorcloth (*boro*) *224*
   hemp 48
   saki-ori 87
   thrift 189, 193
   traditional/modern dress *21*
Japanese maple (*Acer palmatum*) 126, *127*
jeans
   blue dye 193
   caring for 141
   organic cotton 85
   repurposing 201
jewel beetles *32*
Jolly, Jean Baptiste 143

jumpers 30, 31
   baby, eco-printed *109*, *150*
   felted lace 200
   into bags 195
   into cardigan/coat 200
   into felted mat 205
   into skirts 199
   into sofa cushions 195
   unravelling 232
   use in quilts 226

K
kantha cloths 39, *80*, 120, 125, 159
kantha quilts 83
Kashmiri shawl darning 159
keratin 144
kilts 106
kimono *122*
kimono silk *195*
Kinross, Jo 85, 132, 160
Kip, sheepdog *42*, *287*
knitting, unravelling 232
knitting wool, organic *58*

L
la Sorsa, Liz 205
lacework *104*, 203, *204*, *233*
'Lake' *179*
laminated textiles 213, 216
Lamsens, Marina *260*
'Landskin' *219*
lanolin 63
Latvia 113
   traditional dress *106*
laundering 136–7, 140–1, 144–53
   while travelling 149
lavender *148*
leaf matter *258*
Leafcutter Designs 85
leather 69
lichen dyes 244, 251
lilies' anthers *238*
line-drying *136*, *146*, *154–5*
linen 45, 46–7, 254
   'Subject to change' *172*, *173*
linseed compresses 46
Linum usitatissimum (flax) 45, 46
little brown dress project 107
Little House in the Big Woods 108

llama 63
logwood (*Haematoxylon campechianum*) 73, 75
look book 190
lycra 84

M

MacDowell, Colin 25
machine washing 147, 149
madder (*Rubia tinctorum*) 242
'Magic' cloth project 170, *171*
Malvaceae family 50
A Manual of Dyeing Receipts 236–7
Maori *moko* 21
mat, from old jumpers 205
materials
    choice of 118–19, 122
    hunting for 132–3
Mauersberger, Christine 174–5
mending 30, 137, 157–61
    depiction on ceramic fragment 29
    fine wool knit *161*
    running stitch *123*
    woollen vestments *158*
    worn spots *146, 150*
merino jersey prints *18, 47*
merino wool 57, 59, 76
metals 72, 76, 238
    found 266
microns 59
Middle East 63
milk 45, 76, 247
Milkymerino 76, *77, 195*
mindfulness 37, 39, 201
mohair 63
mordanting 246
mordants 245–6
    collecting protein mordant 247
    defined 256
    making 257–8
    'pot as mordant' 266, 275
    stains as 193
    toxic waste 236
Morris, William 89
Morsbags 194, *194*
Morsman, Claire 194
moths 149, 153, 160
mummy cloths 54
Munger, Madeleine 160
Musa sp. (banana) 53

N

Napier, James 236
natural fibres 29, 32, 44
    alpaca 63
    bamboo 55
    banana 45, 53
    bargasse 69
    chemicals on 87
    cotton *see* cotton
    derivation 45
    goat 63
    harakeke 45, 47, *47*
    leather 69
    linen 45, 46–7, 254
    nettle 34, 45, 54–5, *54*
    ramie 54, 254
    silk *see* silk
    undyed cloth *117*
    wool *see* wool
necklaces 231
needle lace, free-form *233*
neem (*Azadirachta indica*) 76
nettle fibre 34, 45, 54–5, *54*
New York 100
New Zealand 115
nightdress, multi-coloured darning *159*
nuclear energy 22
nylon 70

O

Oakes, Amy *86*
obsessions 99
obsolescence 25
ochre 118, 128
off-gassing 84
oil 43, 87, 90
onion bags 218
onionskin dyeing 252–5, *252, 253, 255,* 256
opportunity shops 94, 101, 120, 131, 132, *193*
orange bags 218
organic, meaning of 42–3
organic chemistry 42–3
over-dyeing 193, 197, *229,* 243
overlaying 199
overlockers 123
Oxydendrum arboreum (sourwood) 266

P

Pardi, Lou 23
Parkes, John 180–*1*
patching 30, 189, 243
patchwork 30, 211
'Path' *183*
patterns, on fabric 72, 82, 254, 266
patterns (paper) 120, 129
    from discarded garments 120–1
    ironing 120
    pre-used 133
Pazyryk felts 35
peach 254
peach juice 245
Perkins, William Henry 236–7
perspiration, as pre-mordant 245
pesticides 27, 91
PET bottle fleece 87
pH, testing for 259
Phoenicians 251
photo exercise 189–90
photosynthesis 22
pinafore
    from shift dress 199
    from shirt 202–3, *202*
Pioneer Women's Hut *226*
'Planeta' (exhibition) *103*
plant dyes 126, 189, 236, 244
    acidity 244
    identifying 250–1
    on linen shirt *173*
    and protein fibres 245
    sources 241, 251
    wild harvesting 251
    and wool 58, 250
plastic bags 194
plastics 87
poisons
    in fibres 52
    in plants 250, 251, 254
pollution, by dye houses 53, 89
polyesters 70, 87, 92
post-mordanting 256
potassium bichromate 237
potato leaves 254
pre-felt 213, 215, 275
pressing *see* ironing
printing block, wooden *73*
Project Alabama 166, 275

protein fibres 144
    dyeing 244, 245, 252
provenance 42–3, 84–5
purple dye 251, 275

Q

quilts
    from clothes of deceased 271
    crazy quilts 225
    felted 227–9
    from old clothes 82, 222–6, *222, 223, 226*
    use of machine *227*

R

rafoogar, of India 159
rag weaving 87, 159, 211
ragbags 205
ragu rugs 30
rainfall patterns 22
ramie (*Boehmeria nivea*) 54, 254
rayon 55
ready-made clothes 25
reconstructions 195, 198–203
recycling 28, 87, 92
'Red blanket wagga' *229*
red wine stains 152
'Redland II Map' *175*
Redmond, Lea 85
religious attire 19
remnant boxes 113, 132
resists 72, 219
retting 47
reverse appliqué 166
rhubarb leaves 259
Ricketts, Rowland III 75
Romance was Born 102
rubber 44, 45
Rubia tinctorum (madder) 242
rugs, cleaning 141
running stitch 176, *177*
Russia 35
rust stains 153

S

Safdie, Moshe 83
saki-ori (rag weaving) 87, 159
saliva 152
salt 152
salvage botany 251
sampler, of stitches *116*
saris 51, *81*, 82–3
scarf, tube-felted 215–19
school sewing classes 115
scouring 214
seawater 247, 248, 256, *257*
Second Nature 60
second-hand clothes 94, 95
    repurposing 189–90
selvedges 120
sericin 144
sericulture 65, 67
sewing circles 115–16, 194, 271
sewing machines 30, *30, 38,* 112–13, 128
'Shawl' *183*
sheep *42,* 57–8
    Chocolate Wool Gotland *61*
    coloured *61,* 62
    cross-breeding 59
    English Leicester *56,* 59, *76*
    in India *51*
    Lincoln 59
    'organic' 62
    shearing 61–2, *76*
    Swaledale 60
sheets
    cotton 50
    ironing 156
    for making toiles 129
    mended *161*
    as quilt backing 226
    sides to middle 30, 46
'Shelter' project 170, *171*
shibori-zome 115, 252
shift dresses, uses 199
shirts
    into pinafores 202–3, *202*
    into skirts 205
    refreshed stained silk *191*
shoe shine 140
shopping 24, 25, 102
shoulder pads 189
shrinkage 119–20, 260
shrouds 270
silk 45, *64,* 65–9, *65,* 254
    cocoons 67
    cold-dyed *75*
    dyeing 250, *257,* 260
    with eucalyptus *263*
    ironing 120
    with paeony petals *243*
    shawl 68
    soy silk 44, 45
    washing 144, 145, 147, 151
    wild 68–9
silkworm eggs 67, 68
Singer sewing machine (1927) 30, *30, 38,* 112–13, 128
skin 29
skirts
    from jeans 201, *201*
    from jumpers 199
    from shirtsleeves 205
    from tablecloth 199
    tattered silk *88*
slavery 52, 274
sleeves
    jumpers, reuse 205
    shirts, reuse 205
    turning 115
sneakers, customised 92
snow, as cleaning agent 140–1, *140*
soaking 146, 147
social responsibility 90
socks 131, 205
    darned 30, *160*
    organic merino 85, *132*
'Path' *183*
    reuse 233
Soetsu Yanagi 82
sofa cushions, from jumpers 195
solar steeping 254, *267*
Sopwith Camel biplane 47
sourwood tree (*Oxydendrum arboreum*) 266
South America 50, 52, 63
soy silk 44, 45
soybeans 246
spidersilk 66
'Spy *boro* field jacket' 166, *167*
St John's wort (*Hypericum perforatum*) 126
stains 189
    covering 243
    embroidering 193
    as mordant 275
    removal 152–3
Stevenson, Robert Louis 39

storage space 108
Story, Holly 172–3
string 35
 dress with lace *230*
 garments 204
 hand-twined silk *231*
string-making 118, 120, 159, 204, 211, 230–2
style of dress 131, 266
'Subject to change' 172, *173*
Sugar Cane jean company 69
sugar-cane waste 69
sunlight 149
sustainability 55, 136, 184
synthetic dyes 87
 effects 72
 pollution source 53
synthetic fabrics 32, 70–1, *70*, 244
 dyeing 252, 260
 identifying 118
 origins in oil 23, 43, 87
 recycling 87, 92

T
ta moko 21
tags, on clothing 22, *50*, *59*, *84*, *85*, *85*
Taihu Lake, China 89
tailor's dummy 129
Tanaka, Chuzaburo 158–9
tannins 69
tartan 106
tattoos 19, *20*, 21
teak tree (*Tectona grandis*) *250*, 251
tensile strength, of fabric 122
textiles 23
 cheap 25
 laminated 213, 216
 origins 34–5
 types 244
 see also fabrics
thrift 37, 189, 193
thrift stores 94, 101, 129, 132
tie dyeing 115
tjap (metal stamp) *73*
toiles 129
Toimata, Rachelle 265
tomato leaves 254
tops, from T-shirts 198, 200
traditions 168

trees 22, 58
trend forecasting 105
Trend Union 105
trousers
 drawstring 131
 under skirts 131
 vintage Japanese hemp *133*
 yoga pants *121*
T-shirts
 cost of making 91
 fair-trade cotton *93*
 into tops 198, 200
 layered *199*
Tuominen, Anu 182–3

U
ultraviolet light 149
undergarments 131
 insulating *50*
uniforms 19, 107
United States 48, 82
Urtica dioica (nettle) 54–5, *54*
Uzbek quilt *222*

V
vegetable dyes 118
Victoria, Queen 106
Victoria Markets, Melbourne 113–14
vinegar 151, 218, 259

W
wabi-sabi 164
Waikanae River 256
wardrobe
 author's 129–30, 189
 eco-friendly 129
 planning 101
 sorting 189–90, 192
 for travelling 101, 129
wardrobes (storage units) 108
washing machines 144, 145, 147, 149
water use 85, 92, 118
waxed linen thread 46
wearable art 83
wedding dresses 123–6, *124–5*, *167*
weeds, garment of *33*
wicking effect, of wool 60

wild silks 68–9
Wilder, Laura Ingalls 108
Willans, Tracy 184–5
windfall cloth *36*
windfall walks 31, 262, *263*, 264
wine
 as mordant 245
 red stains 152
Wirrealpa Station 244
wool 45, 57–62, *57*, 212–13, 254
 bonding qualities 213
 as building insulation 60
 classing 59
 drying 151
 dyeing 250, 252, 260
 and eucalyptus 258
 fire resistance 58, 59, 120, 212
 memory for shape 151, 212
 natural colours *58*, *62*
 shrinkage 119, 260
 superwash process 212
 thermal properties 60
 thickness 59
 washing 144, 145, 146–7, 151
 wicking effect 60
wool sliver 212–15
World War II 86, 116, 136, 140

Y
'yerba buena' *249*
yoga pants *121*

Z
zippers, pre-used 133

Photo captions and credits

All photographs are by the author unless stated below.

page 22: label courtesy Lea Redmond
page 29: photo courtesy Petrus Spronk
page 60: Swaledale sheep photo courtesy Second Nature
pages 61 (Gotland sheep) and 62 (naturally coloured wool samples): courtesy Hamish Black of Chocolate Wool NZ
page 74: photo by Rowland Rickets III
page 85: jeans image courtesy Denham the Jeanmaker
page 98: photographer unknown
page 99: photo by Jenni Worth
page 102: Romance was Born photo courtesy University of Technology, Sydney
page 106: photographer unknown
page 112: photo by Jenni Worth
pages 138–139: pieced cloth squares air-drying after being dyed
pages 154–155: the 'Hill's Hoist', one of South Australia's most famous contraptions
page 158: the woollen vestments of St Francis, Assisi, S Francesco, Sacro Convento, Chapter House. Photo: akg-images/Gerhard Ruf
page 162: stitched work by Roz Hawker (photo by India Flint)
page 165: (main) Joost van de Brug; (top and bottom right): Valeria D'Agostino
page 167: courtesy Natalie Chanin – Alabama Chanin
page 169: Spy '*Boro*' jacket image courtesy www.denhamthejeanmaker.com and photographer Ali Kirby @ Denham
page 171: both images courtesy Jude Hill
page 173: 'Subject to change' photo by Victor France
page 175: Christine Mauersberger, 'Redland II Map', 2009. Vintage table linen, embroidery floss. Mounted on felted wool, 45 cm x 45 cm. Photo: Fuchs and Kasparek
page 179: Dorothy Caldwell, 'Lake', 2008. Dyed and printed cotton with stitching, 46 x 46 cm. Photo courtesy the artist
page 181: John Parkes, 'Genes/Jean's', 2009. Recycled cotton cloth (pillowcases and pyjamas), cotton and linen thread, cotton yarn, hand (prick and running) stitch, 164 x 48 cm. Photo courtesy the artist
page 183: Anu Tuominen images courtesy the artist
pages 206–207: a wool and silk blanket (feltquilt), heavily stitched and dyed using *Eucalyptus cinerea* and *E. crenulata*
page 214: photographs by Natasha Milne, © Murdoch Books
pages 220–221: Odd pieces of felt and woven cloth (cotton muslin and silk scraps) hand tacked and machine stitched together prior to fulling
pages 272–273: a collection of randomly pieced plant-dyed fabrics photographed at a workshop in Brisbane

About the author

Designer, artist, writer and sheep farmer India Flint was born in Melbourne and has lived in diverse locations from the Andamooka opal fields to rural Australia and metropolitan Montreal. This gypsy life enriches a textile practice embracing art, theatre and fashion. Presently based on a rural property on the eastern escarpment of the Mount Lofty Ranges, South Australia, she is known for the development of the highly distinctive eco-print, an ecologically sustainable plant-based printing process giving brilliant colour to cloth.

India has been working with plant dyes for more than twenty-five years. Her work is represented in collections and museums in Australia, Latvia and Germany. She produces and sporadically exhibits a range of hand-worked salvaged clothing under the label 'prophet of bloom', as well as designing and making plant-dyed costumes for theatre and dance.

Acknowledgments

At the conclusion of such a project there are always multitudes of people to offer bouquets for all sorts of useful contributions. Some of them are listed here.

To Diana Hill, my publisher at Murdoch Books, for tolerating my shortcomings and shadow-dancing, and to Janine Flew for insightful eagle-eyed editing.

To Toyoko Sugiwaka for turning my text and picture files into a beautiful book.

To my family, friends, skinsisters and students for story-telling, sewing circles, cups of tea, morsels of delicious chocolate (an essential daily vitamin) and the occasional margarita.

To Marion Gorr of Beautiful Silks, Sara Victoria, Hansie Armour of Milkymerino, and Beatrice Kuyumgian-Rankin of the Hemp Gallery for supplying indispensable samples. To the generous artists and makers who have permitted publication of their work in these pages.

To Jo Kinross for knitting my socks and Madeleine Munger for darning them. To Faye Cumming for handing me a lovely linen shirt and allowing me to experiment on it with scissors.

To my mother Arija, aunt Mara, grandmother Berta and great-aunts Rose and Ilse for making sure I knew which way was up. To my children for keeping my feet firmly on the ground. To my lap-cat and studio assistant Martha for providing amusing diversion from time to time, and the unforgettable Kip, now residing with the Dogs Above.

I thank you all.

And of course any errors are mine.

Published in 2011 by Murdoch Books Pty Limited

Murdoch Books Australia
Pier 8/9
23 Hickson Road
Millers Point NSW 2000
Phone: +61 (0) 2 8220 2000
Fax: +61 (0) 2 8220 2558
www.murdochbooks.com.au

Murdoch Books UK Limited
Erico House, 6th Floor
93–99 Upper Richmond Road
Putney, London SW15 2TG
Phone: +44 (0) 20 8785 5995
Fax: +44 (0) 20 8785 5985
www.murdochbooks.co.uk

Publisher: Diana Hill
Project manager and editor: Janine Flew
Designer: Toyoko Sugiwaka
Production: Alexandra Gonzalez

Text copyright © India Flint 2011
Photography (except where stated on page 284) copyright © India Flint 2011
Design copyright © Murdoch Books Pty Limited 2011

All rights reserved. No part of this publication may be reproduced, stored in a retrieval system or transmitted in any form or by any means, electronic, mechanical, photocopying, recording or otherwise, without the prior written permission of the publisher.

National Library of Australia Cataloguing-in-Publication entry
Author: Flint, India.
Title: Second skin : choosing and caring for textiles and clothing/India Flint
ISBN: 978-1-74196-721-0 (hbk.)
Notes: Includes bibliographical references and index.
Subjects: Clothing and dress--Environmental aspects.
Textile fabrics--Environmental aspects.
Textile fabrics--Recycling.
Dewey Number: 646.3
A catalogue record for this book is available from the British Library.

Printed by 1010 Printing International Limited, China.

This book is printed on 100% recycled paper (both white and brown papers) using soya-based inks.